Remote Sensing: Data Analysis and Image Processing

Remote Sensing: Data Analysis and Image Processing

Jaxon Parry

MURPHY & MOORE

www.murphy-moorepublishing.com

Published by Murphy & Moore Publishing,
1 Rockefeller Plaza,
New York City, NY 10020, USA
www.murphy-moorepublishing.com

Remote Sensing: Data Analysis and Image Processing
Jaxon Parry

© 2022 Murphy & Moore Publishing

International Standard Book Number: 978-1-63987-489-7 (Hardback)

Cataloging-in-Publication Data

Remote sensing : data analysis and image processing / Jaxon Parry.
        p. cm.
Includes bibliographical references and index.
ISBN 978-1-63987-489-7
1. Remote sensing. 2. Aerial photogrammetry. 3. Aerospace telemetry.
4. Detectors. I. Parry, Jaxon.
G70.4 .R46 2022
621.367 8--dc23

Table of Contents

Preface

This book aims to help a broader range of students by exploring a wide variety of significant topics related to this discipline. It will help students in achieving a higher level of understanding of the subject and excel in their respective fields. This book would not have been possible without the unwavered support of my senior professors who took out the time to provide me feedback and help me with the process. I would also like to thank my family for their patience and support.

The information gained about an object or phenomenon without making any physical contact with that object is known as remote sensing. This information has four characteristics. They are spatial, temporal, spectral and radiometric resolution. This domain can be divided into two types; Active and passive remote sensing. Active remote sensing is the reflection of signal emitted by a satellite that is indentified by the sensor. Passive remote sensing is the reflection of the light of the sun that is identified by the sensor. This field is used in various other fields such as land surveying, ecology, meteorology, oceanography, hydrology and geography. It also has various commercial, military, planning, intelligence and humanitarian applications. Weather forecasting and reports on climate change are some of the other areas where this discipline finds its application. The book aims to shed light on some of the unexplored aspects of this discipline. It also outlines the processes and applications of remote sensing in detail. It will serve as a valuable source of reference for those interested in this field.

A brief overview of the book contents is provided below:

Chapter – What is Remote Sensing?

Remote sensing is the science of detecting, acquiring and monitoring physical characteristics of an area by measuring its reflected and emitted radiation at a distance or aircraft-based sensor technologies. Active remote sensing, passive remote sensing, optical remote sensing, microwave remote sensing, airborne sensing, etc. are a few of its types. The topics elaborated in this chapter will help in gaining a better perspective of remote sensing.

Chapter – Satellite Remote Sensing

Satellite remote sensing refers to the use of satellites for the estimation of geophysical parameters from the electromagnetic radiations emitted from Earth. IKONOS, SatNav, Envisat, SPOT, SCIAMACHY, etc. are some of its examples. This chapter discusses these types and related aspects of satellite remote sensing in detail.

Chapter – Geographic Information System and Global Positioning System

Geographic information system deals with capturing, storing, manipulating and managing all types of spatial and geographical data. Global positioning system is a navigation system that consists of a network of 24 satellites for determining the ground position of an object. This chapter has been carefully written to provide an easy understanding of geographic information system and global positioning system.

Chapter – Radar and its Types

Radar is referred to the detection system that determines the range, angle and location of objects through the use of radio waves. A few of its types are weather radar, Doppler radar, synthetic-aperture radar, imaging radar, etc. This chapter closely examines the different types of radar to provide an extensive understanding of the subject.

Chapter – Remote Sensing Software

There are many remote sensing software that are globally used such as ERDAS Imagine, PCI Geomatica, TNTmips, IDRISI, Dragon, Google Earth, Opticks, Orfeo toolbox, RemoteView, etc. This chapter sheds light on the different remote sensing software to provide an in-depth understanding of the subject.

Chapter – Applications of Remote Sensing

There are a wide range of applications of remote sensing which include detection and monitoring of marine pollution, biodiversity conservation, environmental resource mapping and modeling, water management, vegetation classification etc. All these diverse applications of remote sensing have been carefully analyzed in this chapter.

Jaxon Parry

1
What is Remote Sensing?

Remote sensing is the science of detecting, acquiring and monitoring physical characteristics of an area by measuring its reflected and emitted radiation at a distance or aircraft-based sensor technologies. Active remote sensing, passive remote sensing, optical remote sensing, microwave remote sensing, airborne sensing, etc. are a few of its types. The topics elaborated in this chapter will help in gaining a better perspective of remote sensing.

"Remote Sensing is the art and science of acquiring information about the earth surface without having any physical contact with it. This is done by sensing and recording of reflected and emitted energy."

In the process of Remote Sensing involves an interaction between the incoming radiation and interest of target. This is done by using imaging and non-imaging system; however the following steps are involved in the process.

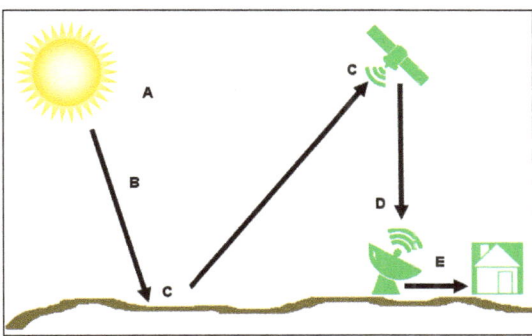

Energy Sources

The first and most important requirement for a Remote Sensing system is an ideal energy source or illumination which provides electromagnetic radiation to the Target interest.

Atmosphere and Radiation

As the energy traveling from its source to Earth surface, it will come in contact with atmosphere when it passes through. This is also happening when the energy from target reflected bake to sensor.

Interaction with the Target and Recording of the Reflected Energy

Once the energy is passed through the atmosphere, it interacts with the target object and depending upon the physical and chemical properties of the Target the energy is reflected or emitted back the Sensor collect and record the Electromagnetic radiation.

Transmission and Ground Level Processing

After the energy sensed it has to be transmitted in the form of electronic signals to the ground stations for processing and generate the output as image (Hard copy/Soft copy).

Interpretation, Analysis and Application

The processed image is interpreted visually and digitally using various interpretation Techniques to extract the information.

The final step is that we are applying the extracted information on various fields of our studies. It may reveal some new information about which the target. The data we gained through Remote Sensing may not be able to collect it through other conventional methods. The applications are infinite.

Electromagnetic Radiation (EMR)

The first and most important component of Remote Sensing is the Energy source to illuminate the Target. The energy is in the form of Electromagnetic Radiation. It is either natural originating from the Sun or earth by emission, or by artificial means. Electromagnetic energy refers to all the energy that moves with the velocity of light in a harmonic wave pattern. EMR consists of an Electrical field and Magnetic field. The electrical field varies magnitude in a direction perpendicular to the direction in which the radiation is travelling and magnetic field oriented to the right angles to the electrical field.

Wave Lengths

A Wave length means the length of one wave cycle or the distance from any position in a cycle to the same position in the next cycle. It is usually represented by Greek letter lambda (λ). Wave length is usually measured in micrometer (mm, 10^{-6}m) and the nanometer (nm, 10^{-9}m).

Frequency

Frequency refers to the number of wave crests passing in a given point in Specific unit of time. It is normally measured in hertz (Hz). Wave length and frequency is related to following formula:

$$v = c / \lambda$$

$$v = \text{frequency} \quad c = \text{Speed of Light} \quad \lambda = \text{Wavelenght}$$

Therefore, these two are inversely related to each other. The shorter wave length, higher the frequency and vice versa.

Electromagnetic Spectrum

The Electromagnetic Spectrum ranges from kilometers to nanometers. These are divided by ranges called Spectral bands. There are several regions in the Electromagnetic spectrum which is useful for Remote Sensing. The following image shows the various regions in EMR.

Region name	Wavelength range	Details
Ultraviolet (UV)	0.30-0.38 μm	Very narrow zone of EMR. It has short wave lengths. Largely scattered by atmospheric particles.
Visible	0.4-0.75 μm	These regions are highly used for Remote Sensing. Comprises of • Violet: 0.4 – 0.446 μm • Blue: 0.446 – 0.500 μm • Green: 0.500 – 0.578 μm • Yellow: 0.578 – 0.592 μm • Orange: 0.592 – 0.620 μm • Red: 0.620 – 0.7 μm Blue, Green and Red are the primary colors in the visible spectrum.
Near Infra red (NIR)	0.75-1.5 μm	Frequently using in Remote Sensing.
Middle Infrared (MIR)	1.5-5 μm	Comprises of • SWIR (1.5-3 μm) • MIR (3.0-5.0 μm) Both the regions are highly using in Remote Sensing.

Thermal Infrared (TIR)	5.0-15.0 µm	Long wavelengths, Much of this energy is comprised of emitted radiation from the Earth.
Microwave	1mm-1m	Longest wavelengths using in Remote Sensing. Using in Both active and Passive mode. These signals can penetrate into clouds and fog.

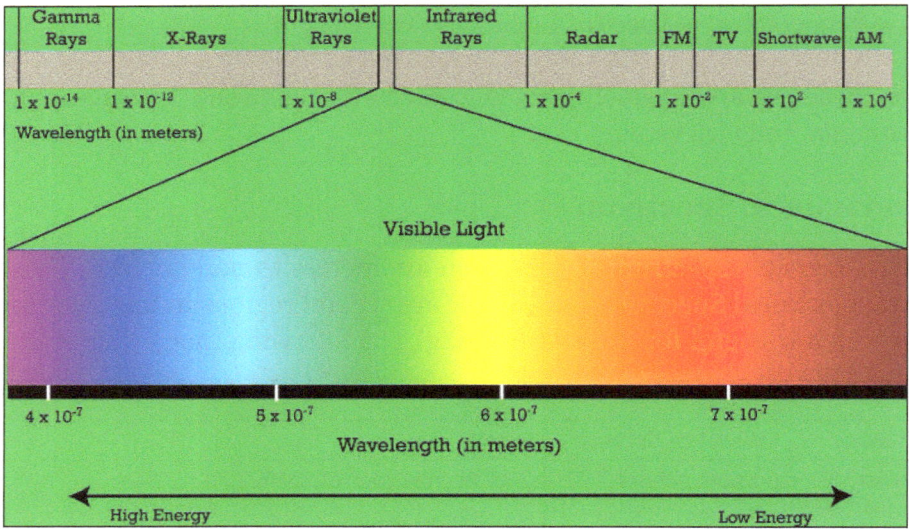

Interactions with Atmosphere

When the incoming solar radiation passes through the atmosphere it may come in contact with atmospheric particles and gases, leads to the mechanisms of scattering and absorption. The gases absorb the Electromagnetic radiation at specific wavelengths called absorption bands. However the high interviewing transmittance regions are often known as Atmospheric Windows.

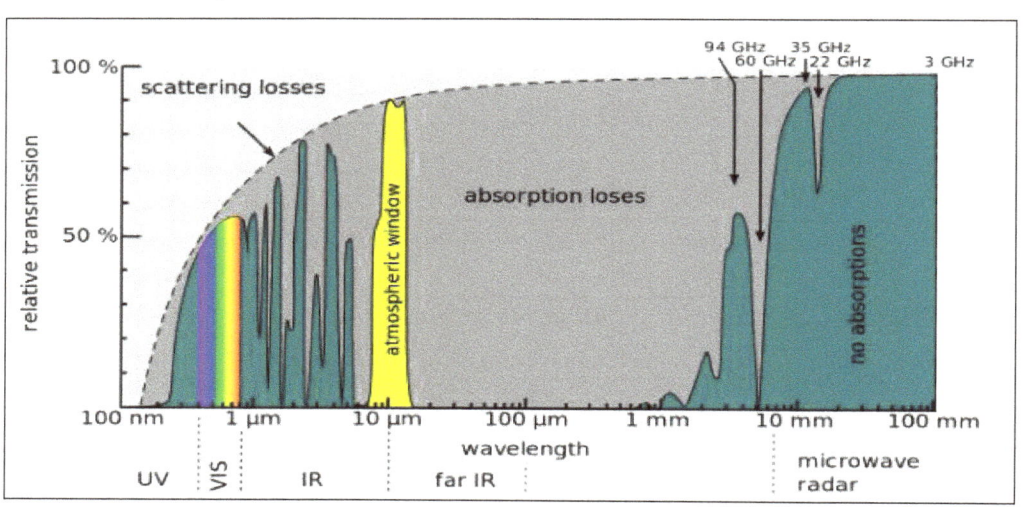

Scattering

When the incoming radiation and light passes through the atmosphere it will affected by the atmospheric particles and this will result the redirection of the light from its original path is known as scattering. The amount of particles present in the atmosphere, wavelength of radiation and the distance of radiation travels through the atmosphere are the major elements which influences the amount of scattering. Following are the major types of scattering.

Rayleigh Scattering

Rayleigh scattering occurs when particles are very small when compared to incoming solar radiation. The small particles of dust, nitrogen and oxygen molecules are causing such type of scattering. The effect of Rayleigh scattering is much more in shorter wavelengths than longer wavelengths. The phenomena of sky appears blue during the day time is because of the Rayleigh scattering the shorter wavelengths of visible spectrum (Blue) scattered more than that of longer wavelength.

Mie Scattering

Unlike Rayleigh scattering, Mie scattering occurs when the incoming solar radiation and the atmospheric particles have the same size. Dust, pollen, smoke and water vapour are causes Mie scattering which tend to affect longer wavelengths than those affected by Rayleigh scattering.

Non Selective Scattering

This occurs when than particles size is much larger than the radiation. Water droplets and dust particles can cause this type of scattering. Non selective scattering gets its name from the fact is that it scattering all the wavelengths equally.

Absorption

Absorption is another main mechanism that happens when the EMR interacts with atmospheric particles and gases. When the EMR passes through the atmosphere the molecules in the atmosphere absorbs energy at various wavelengths. Ozone, CO_2 and water vapour are the main causes of energy absorption:

- Ozone absorbs the harmful ultraviolet rays of incoming solar radiation. This also causes to avoid the ultraviolet rays from Remote Sensing sensors.

- Carbon dioxide is referred as the green house gas. CO_2 tends to absorb the far infrared portion of the spectrum and traps inside which causes the heat inside the atmosphere.

- Water vapour in the atmosphere absorbs the incoming long wave infrared and short wave microwave radiation (between 22 µm and 1m).The presence of water vapour in the atmosphere varies place to place.

Passive and Active Remote Sensing

There are two types of Remote Sensing Systems namely Active and Passive sensing. Passive sensing means the sensor uses Sun's energy as the source of illumination and active sensing means the sensor emitting the energy to the target and collecting the reflected energy. Some examples of active sensors are fluorosensor and Synthetic Aperture Radar (SAR).

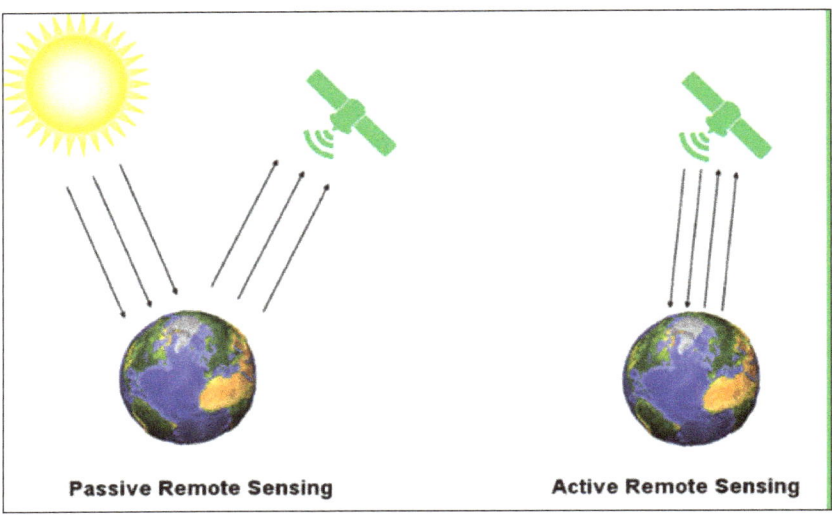

Passive Remote Sensing   **Active Remote Sensing**

The main disadvantage of passive sensors is that they can collect or detect objects in the day time only because sun's illumination is not there at night, however they can record the naturally emitted energy like Thermal infrared. On the other hand Active sensor gives own energy for illumination so it enables to detect and record the images at any time. They are weather independent also; artificial microwaves can penetrate clouds, light and shadow. But Passive sensors are not weather independent. Radar signals can penetrate into vegetation and soil and even can give you the surface information at mm to m depth level at the same time major disadvantage is that radar signals do not contain any spectral characters while Passive Remote Sensing signals have spectral characters. Unlike active sensors passive sensor have the ability to produce fine resolution image. Active Remote sensors are cost intensive also when compared to passive sensor.

Sensors and Satellites

It is the responsibility of the sensor to detect and capture the emitted or reflected energy from the target. In order to capture the energy, the sensor must reside on a stable surface called Platforms. Different types of sensors are there on the basis of Platforms.

Sensors and Types

To collect the reflected or emitted energy the sensor must reside above the target. It may be ground based (lies within the Earth's surface). Aircraft based and Satellite based (lies outside of the Earth's surface).

Ground based sensors are using to record to collect the detailed information about the Earth surface. The data collected by ground based sensor is used to compare the data collected by other sensors like satellite based or to understand the surface features more detailed. One advantage is that the atmospheric disturbances are absent in this type of sensing. The sensor may place on a ladder, scaffolding or crane etc.

Aircrafts are the major platforms in Aircraft based sensors. Helicopters are also occasionally using. Aircrafts are using to collect the detailed information about the Earth surface at any time. Major disadvantage is that it is not possible to fly aircrafts at bad weather.

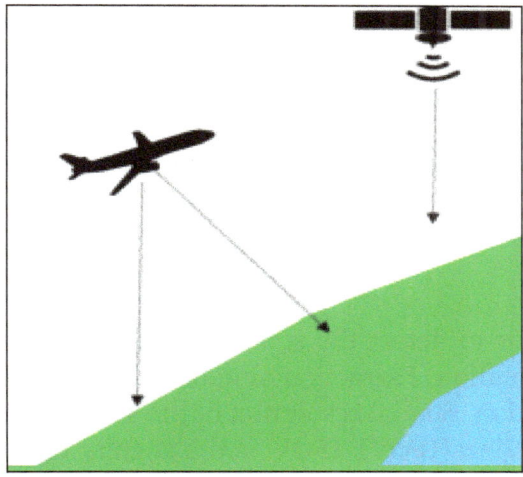

In the space remote sensing are conducted from either using Space shuttle or Satellites. These kinds of Satellites are often known as Remote Sensing satellites. Satellite based sensors are the most efficient and wide spread in the world. These can give a global level coverage but it is costly. Cost is often an important factor while choosing the sensor.

Satellite Characteristics and Resolution Properties

Satellites have some characteristics that makes this is a unique platform in Remote Sensing sensors.

Orbits

The path which is followed by satellites is referred as Orbit. Orbits vary along with their altitude and their orientation and rotation related to the Earth. The orbital change is also influences the satellite's view of earth portion.

To monitor a particular place regularly in the earth surface we are using geostationary orbits. The satellites following the same orbit are called geostationary satellites. This orbit has a very high altitude of 36000KM approximately. The satellites revolving through the orbit have a speed match with the rotation of Earth. This allows monitoring a particular place continuously.

Sun synchronous orbit or polar orbit: some satellites are designed to follow a north south (pole to pole) direction of orbit called Polar orbit. These satellites cover the local time (regularly), referred as Sun synchronous satellites. At any given latitude the, the position of the sun in the sky as the satellite passes overhead will be the same within the same season. This enables the consistent illumination conditions when acquiring images of specific season over successive years.

Resolution Properties

Some instruments taking images from very high altitude and it may give a global coverage or country level coverage. It is not possible to extract information about your home from the global level coverage. In order to extract information we have to design a new sensor to provide Centimeter level information. While designing the sensor or using the output image we have to look up about resolution. There are four distinct types of resolutions. They are:

Spatial Resolution

The ability of the sensor to detect the smallest single object in the Earth surface is referred as Spatial Resolution. Extraction of details from the image is highly depends upon the spatial resolution. Spatial resolution of the sensor is highly depends on their Instantaneous Field of View (IFOV). IFOV is the angular cone of visibility of the sensor. It determines the area of the Earth's surface which is "seen" from a given altitude at particular moment of time. The size of the area viewed is determined by multiplying IFOV by the distance from the ground to the sensor.

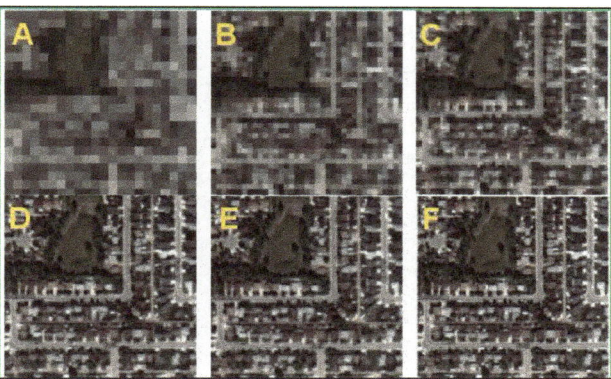

Spectral Resolution

Simply we can define as the ability of the sensor to record the information on a particular spectral range is called Spectral Resolution or the ability of the sensor to define fine wavelengths. Spectral resolution is highly important while designing the sensor. It defines the nature of the study. Different objects have different spectral signature.

Temporal Resolution

Like other resolution properties temporal resolution is also important in Remote Sensing studies. Temporal resolution means the revisit of the satellite over the particular area at same interval of time. The temporal resolutions of the satellites are usually several days which enable to capture the images of an area regularly and can monitor well. The images collected over a given interval are called multi-temporal data, which is the main advantage.

Multi Spectral and Thermal Imaging

Remote Sensing instruments are acquiring the data by scanning the Earth surface. Different types of scanning systems are using in Remote Sensing for acquiring the data.

Multi Spectral Scanning

A scanning system collecting data over a various wavelengths but not continuous is referred as Multi spectral scanning system. It is the most commonly using scanning system in the world. There are two methods are deployed in the multi spectral scanning system for acquiring information about the Earth surface. They are follows:

Across Track Scanning

Across Track scanning scans the surface in a series of lines. These lines are perpendicular to the direction of the motion of the sensor. Each line is scanned from one side of the sensor to another side. This is done by using rotating mirror. This type scanning is also known as whisk broom scanning.

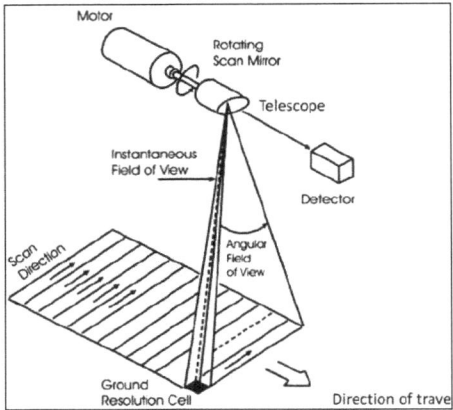

Along Track Scanning

This type scanning uses a forward motion of the sensor to detect the successive scan lines and these lines are perpendicular to the flight direction. These systems are also referred as push broom scanning system. Instead of rotating mirror here using a linear array of detectors located I the focal plane of the sensor.

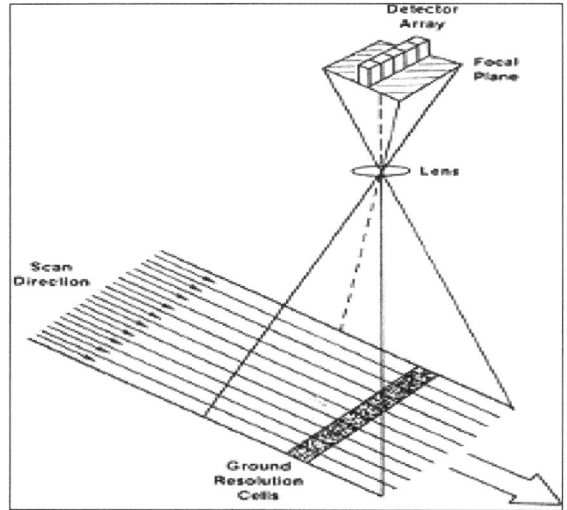

Thermal Imaging

Multi spectral systems usually using reflected energy for acquiring the information, while unlike them Thermal imaging is using emitted energy like thermal infrared spectrum (3 μm to 15 μm) for acquiring data. Unlike multi spectral system Thermal sensors using photo deters sensitive to direct contact of photons on their surface, to detect emitted thermal radiation. This detectors are cooled (Temperature below 0 °C) for avoiding their own emission. These sensors are measuring the surface temperature of the target. Thermal Imaging is typically using the across track scanning system. Thermal sensors normally have large IFOV to ensure enough energy reaching to the sensor to make reliable measurements.

Image Processing and Analysis

Interpretation and analysis of remotely sensed image involves the identification of the target it may be an object or feature in the Earth surface. We can extract the information and analysis either by manual method or aided by computer. Before making computer aided techniques, it is better to extract and identify feature manually called Visual interpretation. Much identification and interpretation of the targets in Remote Sensing are done by visual interpretation i.e. human interpreter. The computer aided techniques called Digital Image processing is simplifying the visual interpretation. It may be used to enhance the data like enhancing the brightness of the data.

Visual Interpretation

Many of the Remote Sensing studies using human interpretation or visual interpretation techniques. How we can do a visual interpretation? There are certain elements are there which makes visual interpretation easier. They are:

- Tone refers to the brightness or color of the image. Tone is an important element in identifying target. Variations in tone enable to identify the target along with other elements like shape, size etc.

- Shape the general form or structure of the individual objects. Shape is the most helpful interpretation technique. Shape of a sharp edge may be an agricultural field rather than a naturally created forest. Linear shape of the road and rail way makes it easy to identify.

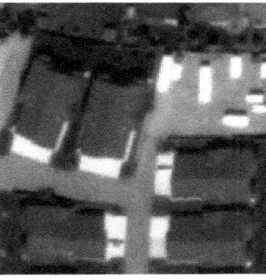

- Size of objects in an image is function of scale. It is important o assess the size of the target relative to other objects in a scene, as well as he absolute size, to aid the interpretation of the target. For example if an interpreter identified an area having larger buildings such as factories or war houses then it may be an industrial region while a clustering of small building indicates they may be a settlement area.

- Pattern means the spatial arrangement of visibly discernible objects in the image. Pattern helps to identify objects having similar characteristics like orchard have evenly spaced tress which may show a similar pattern which makes identification easier. Regularly spaced villas are another example of same pattern.

- Texture is the arrangement and frequency of tonal variations in particular area of an image. Several types of textures are there like rough texture and smooth texture. Smooth texture will have uniform, even surface and little tonal variations. Grassland is a typical example for smooth texture. While rough texture may change tonal variations abruptly. Texture is one of the most important elements for distinguishing features in Radar imagery.

- Shadow can help the interpretation easier at the same time it can reduce inter-pretation. It can provide the idea about the relative relief of the target which makes identification easier. While if the shadow falls on an object it will hide that object and it makes much difficult to interpreter.

- Association means simply the relationship between nearby objects or feature to the target. One would be expected to associate other like play ground in the schools. This makes identification easier.

Digital Image Processing

Today most of the remote sensing data is recording in digital formats. These data is storing in various formats like BSQ (Band sequential), BIP (Band Interleaved by Line) and BIP (Band interleaved by Pixel) and so on and these images are undergoing var-ious image processing steps. Digital image processing involves numerous procedures including correction of data, image enhancements for better visualization, fully auto-mated classification or interpreter aided computer classification of the images and so on. Minimum facilities are required for image processing system is that a computer with necessary hardware which meets the requirement for image processing and a good software package and Remote Sensed image (in digital format).Several commercial software packages are available in the market. All these systems are collectively known as Image analysis system. The major image processing functions can bring under three main heading as follows.

Preprocessing

Preprocessing is usually done prior to the main analysis. These are sometimes referred as image rectification and restoration. Preprocessing is indented to correct the platform, sensor radiometric and geometric distortions in the data. Radiometric corrections are done to stabilize the illumination variation and viewing geometry. These are happening due to distortion in platforms and sensors. It can be corrected by modeling the geometry relationship and distance between the areas of the Earth surface imaged. Noise of the image may be due to the errors or distortions occurring while recording the data. Common forms of noise include systematic striping or banding and dropped line it should be corrected before the enhancements are doing.

Enhancements and Classification

Enhancement techniques are applied to digital imager prior to visual image interpretation. This optimizes the imagery for visual interpretation purpose. Even we have done radiometric corrections in the image to reduce the noise and other atmospheric disturbances still may be in the image and is not optimized for visual interpretation. If the image have large variations in spectral responses from variety of targets, simple radiometric correction can't optimize he image in its all portions.For each application and each image a custom enhancements are needed. To understand enhancement techniques it is better to understand the term image histogram. A histogram is a graphical representation of the brightness values that comprise an image. The brightness values are plotted in X axis and the frequency of occurrence of each of these values in the image shown on the Y axis.

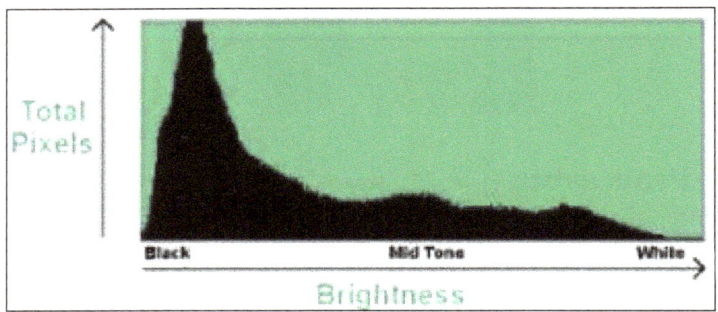

Image Histogram.

Contrast enhancement is an enhancement technique normally involving changing the original values so that more of the available range is used, thereby increasing the contrast between targets and their backgrounds. Linear Contrast Stretching is the simple type of enhancement techniques involving identification of minimum and maximum brightness values in the image and applying to stretch this range to fill the full range. Histogram equalization is another technique. This is a non linear stretch. In this method the DN values are redistributed on the basis of their frequency. More efficient gray tones are assigned to the frequently occurring DN values of the histogram. Spatial filtering is a technique which involves noise reduction (Low pass filter) and edge enhancement (High pass filter).

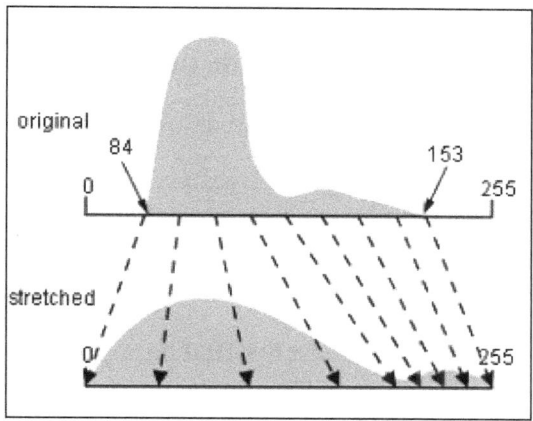

Linear Contrast Stretching.

Digital Image classification the computer classifies the image by DN values of each pixel on the basis of spectral responses. It may be either fully automated referred as Unsupervised classification. Here computer classifies the image and interpreter says only how many classes (Building, forest, agricultural lands and so on) he wants and based on that the system generates the classified image. Supervised classification is another method involves the interpreter have regulations on the classification. Interpreter identifies the pixels of each class and assigned to computer and based on that the computer classifies the image.

Advantages of Remote Sensing Technology

- Large area coverage: Remote sensing allows coverage of very large areas which enables regional surveys on a variety of themes and identification of extremely large features.

- Remote sensing allows repetitive coverage which comes in handy when collecting data on dynamic themes such as water, agricultural fields and so on.

- Remote sensing allows for easy collection of data over a variety of scales and resolutions.

- A single image captured through remote sensing can be analyzed and interpreted for use in various applications and purposes. There is no limitation on the extent of information that can be gathered from a single remotely sensed image.

- Remotely sensed data can easily be processed and analyzed fast using a computer and the data utilized for various purposes.

- Remote sensing is unobstructive especially if the sensor is passively recording the electromagnetic energy reflected from or emitted by the phenomena of interest. This means that passive remote sensing does not disturb the object or the area of interest.

- Data collected through remote sensing is analyzed at the laboratory which minimizes the work that needs to be done on the field.

- Remote sensing allows for map revision at a small to medium scale which makes it a bit cheaper and faster.

- Color composite can be obtained or produced from three separate band images which ensure the details of the area are far much more defined than when only a single band image or aerial photograph is being reproduced.

- It is easier to locate floods or forest fire that has spread over a large region which makes it easier to plan a rescue mission easily and fast.

- Remote sensing is a relatively cheap and constructive method reconstructing a base map in the absence of detailed land survey methods.

Disadvantages of Remote Sensing

- Remote sensing is a fairly expensive method of analysis especially when measuring or analyzing smaller areas.

- Remote sensing requires a special kind of training to analyze the images. It is therefore expensive in the long run to use remote sensing technology since extra training must be accorded to the users of the technology.

- It is expensive to analyze repetitive photographs if there is need to analyze different aspects of the photography features.

- It is humans who select what sensor needs to be used to collect the data, specify the resolution of the data and calibration of the sensor, select the platform that will carry the sensor and determine when the data will be collected. Because of this, it is easier to introduce human error in this kind of analysis.

- Powerful active remote sensing systems such as radars that emit their own electromagnetic radiation can be intrusive and affect the phenomenon being investigated.

- The instruments used in remote sensing may sometimes be un-calibrated which may lead to un-calibrated remote sensing data.

- Sometimes different phenomena being analyzed may look the same during measurement which may lead to classification error.

- The image being analyzed may sometimes be interfered by other phenomena that are not being measured and this should also be accounted for during analysis.

- Remote sensing technology is sometimes oversold to the point where it feels like it is a panacea that will provide all the solution and information for conducting physical, biological or scientific research.

- The information provided by remote sensing data may not be complete and may be temporary.

- Sometimes large scale engineering maps cannot be prepared from satellite data which makes remote sensing data collection incomplete.

Elements of Remote Sensing Systems

- Remote Sensing Systems: These systems consist of a number of elements. The elements range from solar radiation to the application of imageries for public problems, through various stages.

- Sun as a Source of Energy: Sun is the prime source of energy to the world. It is the resource for all activities of the life forms.

- Emission of Sun's Energy: Solar energy is emitted into the space. The emission is in the form of various electromagnetic waves. It consists of gamma rays to radio waves (short wave length). This band of rays is called the 'electromagnetic radiation' (EMR).

- Interaction of Solar Energy with Atmospheric Elements: When the solar energy passes through the atmosphere, many elements await to meet the energy. A portion of the 'electromagnetic radiation' is absorbed by carbon di oxide, ozone, moisture and dust and reflected back. So, the balance of electromagnetic radiation reaches the earth's surface as sunlight.

- Interaction of Sunlight with Terrestrial Features: Electromagnetic waves in sunlight have different wave lengths. A number of bands can be identified based on the wave lengths. These bands of radiation fall on the objects of the earth and get reflected, differently by different objects. The reflectance varies according to the wave length. Through such reflectance, various wave lengths help in remote sensing to identify various elements over the earth. Thus, the spectral reflectances from the earth, in fact earth objects, are of many thousand types.

- Terrestrial Radiation of the Earth's Element: The solar energy, along with the energy already stored in the elements on the earth, are radiated back into the atmoshpere. Any object, with a temperature of about oo K (2730 C) will emit energy. Thus, all objects over the earth have temperatures above oo K and therefore emit energy at varying levels.

- Collection of Information/Data: The energy thus reflected and emitted by the earth's features are recorded by cameras and sensors fitted onto the various platforms. The cameras record the energy in films and the sensors convert the energy into electrical signals and send them to the earth's receiving stations.

- Data Acquisition by the Earth Station: The electrical pulses from the remote sensors are converted into 'digital' numbers. Each point or picture element gets different (pixel) digital numbers. Thus, a satellite image is composed different digital values or pixels.

Data Acquisition and Integration

Remote sensing is the measurement of the acquisition of data about the Earth's surface without contact with it. This is done by sensing and recording reflected or emitted electromagnetic radiation. Remote sensing involves analyzing and applying that information. The process involves the following elements:

- Energy source: The first requirement for remote sensing is an energy source which provides electromagnetic energy.

- Radiation and the atmosphere: As the energy travels from its source to the target, it will come in contact with and interact with the atmosphere it passes through. This interaction may take place a second time (active remote sensing) as the energy travels from the target to the sensor.

- Interaction with the target: Once the energy makes its way to the target through the atmosphere, it interacts with the target depending on the properties of both the target and the radiation.

- Recording of energy by the sensor: After the energy has been reflected by, or emitted from the target, we require a sensor (remote - not in contact with the target) to detect and record the electromagnetic radiation.

- Transmission, reception, and processing: The energy recorded by the sensor has to be transmitted, often in electronic form, to a receiving and processing station where the data are processed into an image (hardcopy and/or digital).

- Interpretation and analysis: The processed image is interpreted, visually and/or digitally, to extract information about the target.

- Application: The final element of the remote sensing process is achieved when we apply the information we have been able to extract from the imagery about the target in order to better understand it, reveal some new information, or assist in solving a particular problem.

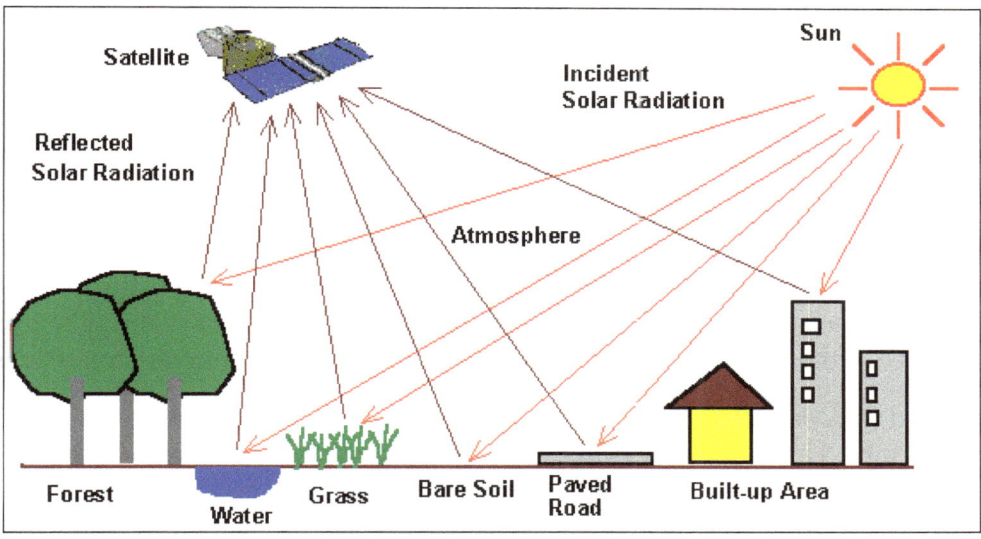

The physical elements of remote sensing process: Energy source, radiation and the atmosphere, interaction with the target, sensing, analysis and application.

Electromagnetic Radiation

The energy to use in remote sensing is in the form of electromagnetic radiation. For understanding remote sensing we need to understand two important characteristics of electromagnetic radiation, which are basic to wave theory. These are the wavelength and frequency. The wavelength is the distance between successive wave crests. Wavelength is usually represented by lambda (λ) and measured in meters or some factor of meters such as nanometers (nm, 10^{-9}m) or micrometers (μm, 10^{-6}m). Frequency refers to the number of cycles of a wave passing a fixed point per unit of time. Frequency is normally measured in hertz (Hz) which is cycle per second. Frequency and wavelength are inversely proportional: the higher the frequency, the shorter the wavelength.

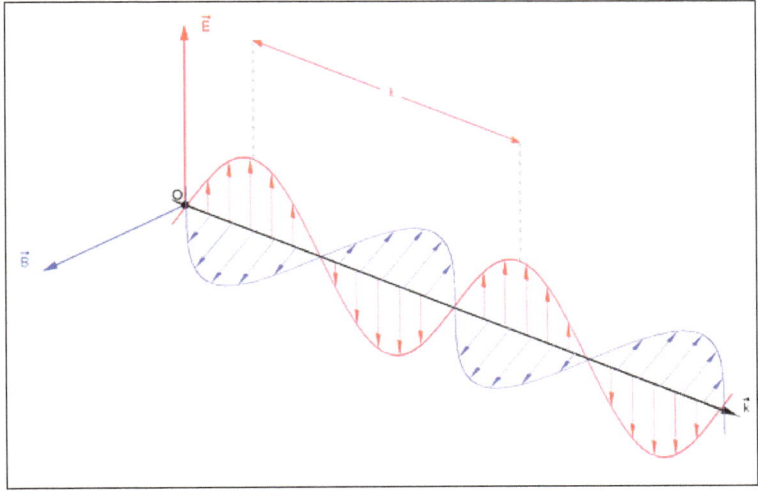

Electromagnetic wave, λ-wavelength.

Electromagnetic Spectrum

The electromagnetic spectrum ranges from the shorter wavelengths (including gamma and x-rays) to the longer wavelengths (including microwaves and broadcast radio waves). There are several regions of the electromagnetic spectrum which are useful for remote sensing.

The visible wavelengths cover a range from approximately 0.4 to 0.7 μm. The light which human eyes can detect is part of the visible spectrum. This is the only portion of the spectrum we can associate with the concept of colors. The primary colors of the light are blue, green and red. Other colors can be made by combining them in various proportions.

The infrared (IR) part of the electromagnetic spectrum covers the range from roughly 0.7 μm to 1 mm. The infrared region can be divided into two categories based on their radiation properties - the reflected IR, and the emitted or thermal IR. Physical processes that are relevant for this range are similar to those for visible light. The reflected IR covers wavelengths from 0.7 μm to 5.0 μm. It can be divided into near- and mid -parts. The thermal IR region is quite different than the visible and reflected IR portions, as this energy is essentially the radiation that is emitted from the Earth's surface in the form of heat. The thermal IR covers wavelengths from approximately 3.0 μm to 100 μm. The Earth emits most strongly in approximately 10 μm.

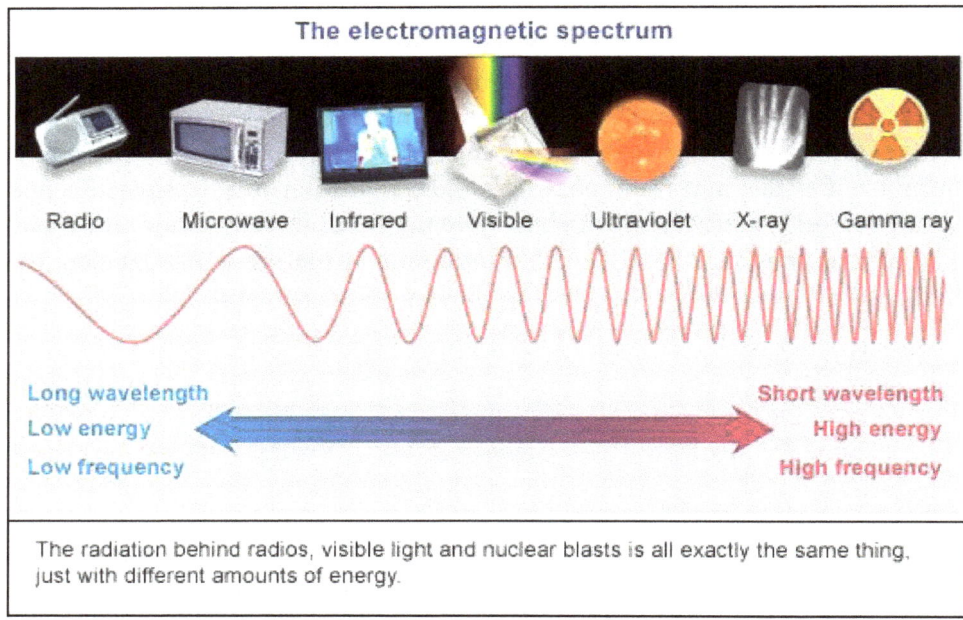

Types of energy level changes associated with different part of electromagnetic spectrum.

The microwave covers region from about 1 mm to 1 m. This covers the longest wavelengths used for remote sensing. The shorter wavelengths have properties similar to the thermal infrared region while the longer wavelengths approach the wavelengths used for radio broadcasts.

Interaction with the Atmosphere

Particles and gases in the atmosphere have effect on remote sensing data and on spectral band selection. These effects are caused by the mechanisms of scattering and absorption. Scattering occurs when radiation is reflected or refracted by particles or large gas molecules present in the atmosphere. Redirection of the electromagnetic radiation depends on several factors including the wavelength of the radiation, the size of particles or gases, and the distance the radiation travels through the atmosphere.

Absorption is the other main mechanism when electromagnetic radiation interacts with the atmosphere and molecules of the atmosphere to absorb energy at various wavelengths. Ozone, carbon dioxide, and water vapour are the three main atmospheric constituents which absorb radiation. These gases absorb electromagnetic energy in very specific regions of the spectrum. There are some regions of the spectrum where radiation is passed through the atmosphere with relatively little attenuation and are useful to remote sensing. Those regions are called atmospheric windows.

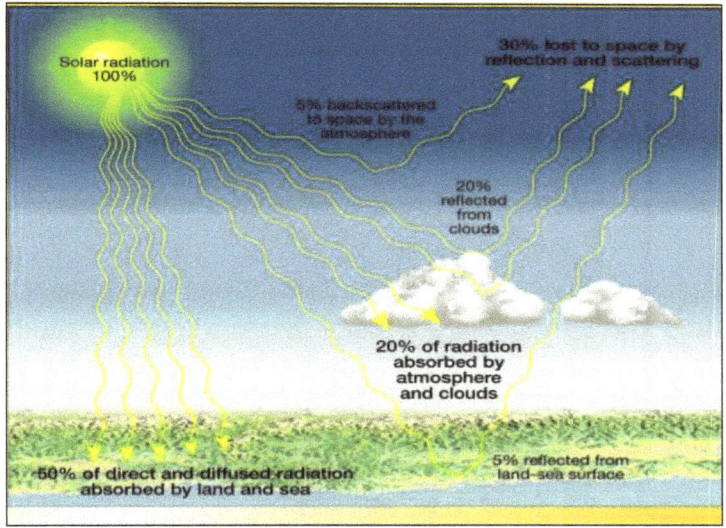

Interaction with the Atmosphere: scattering, absorption.

Interaction with the Target

Radiation passed through the atmosphere interact with the Earth's surface. There are three forms of interaction: absorption, transmission and reflection. In remote sensing, we are most interested in measuring the radiation reflected from targets. We refer to two types of reflection, which represent the two extreme ends of the way in which energy is reflected from a target: specular and diffuse reflection. The interaction with the surface depends on the wavelength of the energy and the material and condition of the surface feature. Different materials reflect and absorb differently the electromagnetic spectrum. The reflectance spectra of a material is a plot of the fraction of radiation reflected as a function of the incident wavelength and serves as a unique signature for the material. The following graph shows the typical reflectance spectra of five materials: clear water, turbid water, two types of soil (dry and wet), and vegetation.

The reflectance of clear water is generally low. The reflectance is maximum at the blue part of the spectrum and decreases as wavelength increases. Turbid water has some sediment suspension which increases the reflectance in the red part of the spectrum. Soil reflectance dependents on its physical and chemical properties. Most important factors determining reflectance are the following: organic matter, moisture content of soil, parent rock, existing colored compounds. In the example shown, the reflectance increases monotonically with increasing wavelength. If moisture content of soil increases, there is a decrease in reflectance. The reflectance of vegetation is low in both the blue and red regions of the visible spectrum, due to absorption by chlorophyll for photosynthesis. It has a peak at the green region which gives rise to the green color of vegetation. In the near infrared (NIR) region, the reflectance is much higher than that in the visible band due to the cellular structure in the leaves. In the mid infrared there are more water absorption regions.

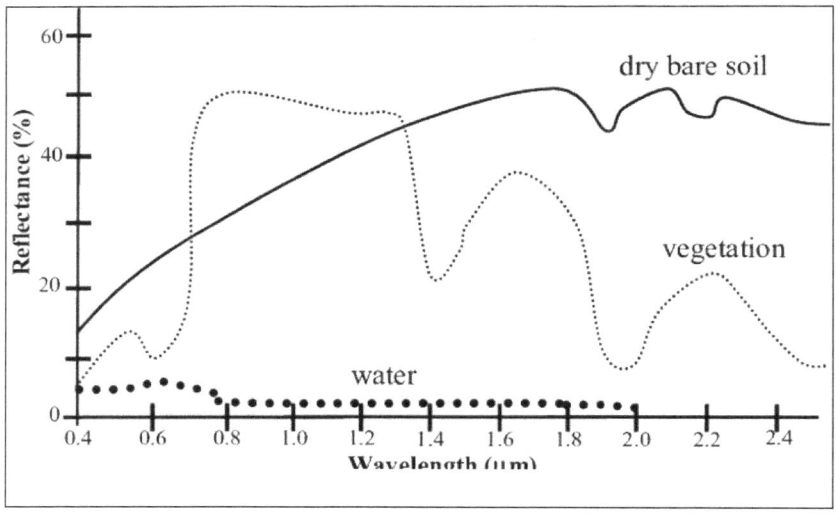

Reflectance of common natural objects: water, soil and vegetation.

Microwave Remote Sensing

Microwave remote sensing is conducted by using microwave radiation and wavelengths from about one centimeter to a few tens of centimeters enables observation in all weather conditions without any restriction by cloud or rain. This is an advantage that is not possible with the visible and/or infrared remote sensing. In addition, microwave remote sensing provides unique information on for example, sea wind and wave direction, which are derived from frequency characteristics, Doppler effect, polarization, back scattering etc. that cannot be observed by visible and infrared sensors. However, the need for sophisticated data analysis is the disadvantage in using microwave remote sensing.

There are two types of microwave remote sensing; active and passive. The active type receives the backscattering which is reflected from the transmitted microwave which is incident on the ground surface.

Synthetic aperture radar (SAR), microwave scatterometers, radar altimeters etc. are active microwave sensors. The passive type receives the microwave radiation emitted from objects on the ground. The microwave radiometer is one of the passive microwave sensors. The process used by the active type, from the transmission by an antenna, to the reception by the antenna is theoretically explained by the radar equation.

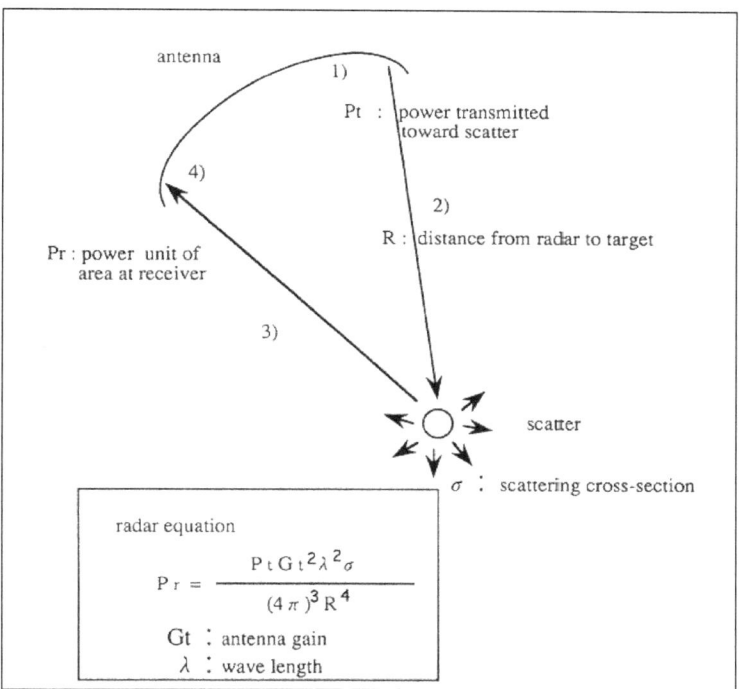

Concept of radar equation.

The process of the passive type is explained using the theory of radiative transfer based on the law of Rayleigh Jeans In both active and passive types, the sensor may be designed considering the optimum frequency needed for the objects to be observed.

In active microwave remote sensing, the characteristics of scattering can be derived from the radar cross section calculated from received power Pr and antenna parameters (A_t, P_t, G_t) and the relationship between them, and the physical characteristics of an object. For example, rainfall can be measured from the relationship between the size of water drops and the intensity of rainfall.

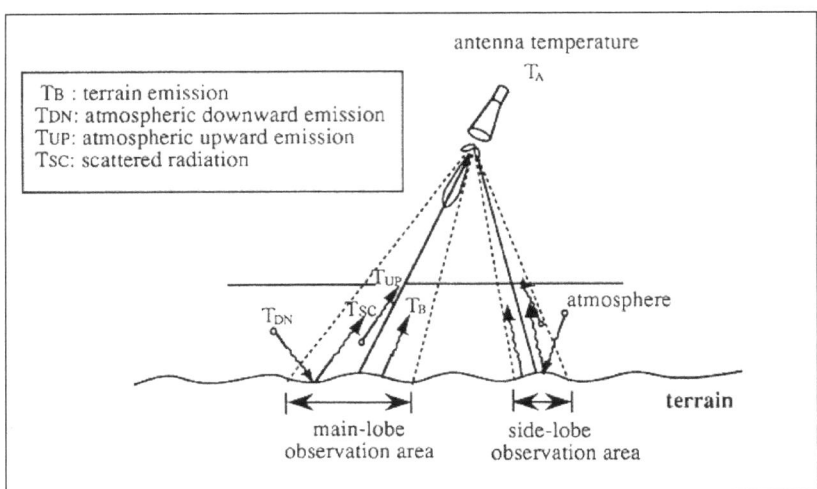

TB : terrain emission
TDN: atmospheric downward emission
TUP: atmospheric upward emission
TSC: scattered radiation

Principal of passive microwave sensor. The apparent temperature represent the energy incident upon the antenna.

In passive microwave remote sensing, the characteristics of an object can be detected from the relationship between the received power and the physical characteristics of the object such as attenuation and radiation characteristics.

Thermal Infrared Remote Sensing

Thermal remote sensing is the branch of remote sensing that deals with the acquisition, processing and interpretation of data acquired primarily in the thermal infrared (TIR) region of the electromagnetic (EM) spectrum. In thermal remote sensing we measure the radiations 'emitted' from the surface of the target, as opposed to optical remote sensing where we measure the radiations 'reflected' by the target under consideration.

It is a well known fact that all natural targets reflect as well as emit radiations. In the TIR region of the EM spectrum, the radiations emitted by the earth due to its thermal state are far more intense than the solar reflected radiations and therefore, sensors operating in this wavelength region primarily detect thermal radiative properties of the ground material. However, very high temperature bodies also emit substantial radiations at shorter wavelengths. As thermal remote sensing deals with the measurement of emitted radiations, for high temperature phenomenon, the realm of thermal remote sensing broadens to encompass not only the TIR but also the short wave infrared (SWIR), near infrared (NIR) and in extreme cases even the visible region of the EM spectrum.

Thermal remote sensing, in principle, is different from remote sensing in the optical and microwave region. In practice, thermal data prove to be complementary to other remote sensing data. Thus, though still not fully explored, thermal remote sensing reserves potentials for a variety of applications.

In thermal remote sensing, radiations emitted by ground objects are measured for temperature estimation. These measurements give the radiant temperature of a body which depends on two factors - kinetic temperature and emissivity.

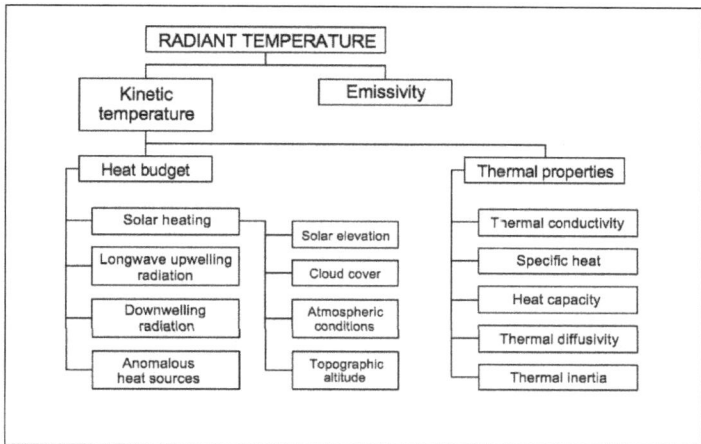

Factors controlling radiant temperature.

Wavelength/Spectral Range

The infrared portion of the electromagnetic spectrum is usually considered to be from 0.7 to 1,000 µm. Within this infrared portion, there are various nomenclatures and little consensus among various groups to define the sub-boundaries. In terrestrial remote sensing the region of 3 to 35 µm is popularly called thermal-infrared. As in all other remote sensing missions, data acquisitions are made only in regions of least spectral absorption known as the atmospheric windows. Within the thermal infrared an excellent atmospheric window lies between 8-14 µm wavelength. Poorer windows lie in 3-5 µm and 17-25 µm. Interpretation of the data in 3-5 µm is complicated due to overlap with solar reflection in day imagery and 17-25 µm region is still not well investigated. Thus 8-14 µm region has been of greatest interest for thermal remote sensing.

Spectral Emissivity and Kinetic Temperature

Thermal remote sensing exploits the fact that everything above absolute zero (0 K or -273.15 °C or −459 °F) emits radiation in the infrared range of the electromagnetic spectrum. How much energy is radiated, and at which wavelengths, depends on the emissivity of the surface and on its kinetic temperature. Emissivity is the emitting ability of a real material compared to that of a black body, and is a spectral property that varies with composition of material and geometric configuration of the surface. Emissivity denoted by epsilon (ε) is a ratio and varies between 0 and 1. For most natural materials, it ranges between 0.7 and 0.95. Kinetic temperature is the surface temperature of a body/ground and is a measure of the amount of heat energy contained in it. It is measured in different units, such as in Kelvin (K); degrees Centigrade (°C); degrees Fahrenheit (°F).

Black Body

Black body is a theoretical object that absorbs and then emits all incident energy at all wavelengths. This means that the emissivity of such an object is by definition 1. Needless to say, such an object is only imaginary and no natural substance is an ideal black body.

Factors affecting the Kinetic Temperature

Factors Affecting the Kinetic Temperature can be categorised in two broad groups - heat energy budget and thermal properties of the materials. Heat energy budget includes factors such as solar heating, longwave upwelling and downwelling radiations, heat transfer at the earth-atmosphere interface and active thermal sources such as fires, volcanoes etc. Thermal properties of material include factors such as thermal conductivity, specific heat, density, heat capacity, thermal diffusivity and thermal inertia of the material.

Radiant Temperature

The radiant temperature (T_R) is the actual temperature obtained in a remote sensing measurement and, as mentioned earlier, depends on actual or kinetic temperature (T_K) of the body and the its emissivity (ε). The total radiations emitted by non black body (natural surfaces) is given by:

$$W = e \cdot s \cdot T_K^4 = s \cdot T_R^4 ,$$

where σ is the Stefan-Boltzmann's constant. This defines the relation between the radiant temperature and kinetic temperature of a body as:

$$T_R = e^{1/4} \cdot T_K$$

From the above equation, and the knowledge that all natural materials are non-black bodies with emissivity less than one, it is clear that the radiant temperature (temperature estimated by remote sensing data) is always less than the actual surface temperature of the body by a factor $e^{1/4}$.

The total amount of radiations emitted by a body can also be estimated using Planck' equation which gives:

$$W_1 = \frac{2p \cdot h \cdot c^2}{1^5} \cdot \left(\frac{1}{e^{h \cdot c / 1 \cdot k \cdot T} - 1} \right) \cdot e_1,$$

where W_1 is the spectral emittance, h is the Planck's constant ($6.62 \cdot 10^{-34}$ js), c is the speed of light ($3 \cdot 10^8$ ms^{-1}), λ is the wavelengtn metres, k is Boltzmann's constant ($1.38 \cdot 10^{-23}$ JK^{-1}), T is the temperature in K and el is the spectral emissivity.

This formula also implies that with the rise in temperature of the ground objects, there is an increase in the intensity of the emitted radiations, with the peak shifting towards shorter wavelengths. Inverting Planck's equation we get:

$$\therefore T = \frac{C_2}{1.1\ln\left(\left[e_1 \cdot C_1 \cdot 1^{-5}\Big/W_1\right]+1\right)},$$

where $C_1 = 2p \cdot h \cdot c = 3.742.10^{-16}$ Wm^{-2} and $C_2 = \dfrac{h.c}{k} = 0.0144$ mK.

Issues

Due to the fundamental difference between remote sensing in the thermal infrared region and the other regions of the EM spectrum, there are some issues peculiar and pertinent for thermal remote sensing. Some of these relate to the mode of acquisition, calibration, radiometric and geometric correction.

Data Acquisition: Modes and Platforms

There are three different aspects which must be considered while talking about the mode of thermal data acquisition. These are.

Active versus Passive Mode

Most of the thermal sensors acquire data passively, i.e. they measure the radiations emitted naturally by the target/ground. Data can also be acquired in the TIR actively deploying laser beams (LiDAR). However, these techniques are not well researched and are only in the infancy.

Broadband versus Multispectral Mode

For the broadband thermal sensing, in general the 8 to 14 µm atmospheric window is utilised. However, some spaceborne thermal sensors such as Landsat Thematic Mapper Band 6 operate in the wavelength range of 10.4 to 12.6 µm to avoid the ozone absorption peak which is located at 9.6 µm. The multispectral thermal channels, such as those in the ASTER platform, are targeted specially for geological applications.

Daytime versus Night-time Acquisition

Thermal data can be acquired during the day and during the night. For some applications it is useful to have data from both the times. However, for many applications night-time or more specifically pre-dawn images are preferred as during this time the effect of differential solar heating is the minimal. The platforms for such data acquisitions range from satellites, aircrafts to ground based scanners.

Spatial Resolution and Geometric Correction

Most thermal sensors have on board recording and calibration systems. Two black bodies (BB) commonly known as BB1 and BB2 are setup which control the radiometric calibration of the acquired data. As the sensors measure emitted radiations, there is also a heating effect and constant cooling of the sensors is required. This poses a physical limit to the measuring capability of the sensors and therefore the spatial resolution of the acquired data. The coarse spatial resolution, specially of satellite borne broad band thermal data poses some additional problems in geometrically registering it to other data, specially when the latter have much higher spatial resolution. Identification of corresponding reliable control points on data sets with such wide differences in spatial resolution is not only difficult but when tried may result in unacceptable transformation results. Alternate approaches of co-registration must be thought of. This may be done by first registering the thermal image to another image with intermediate spatial resolution and in the next step to the target high resolution image.

Applications

Thermal property of a material is representative of upper several centimetres of the surface. As in thermal remote sensing we measure the emitted radiations, it proves to be complementary to other remote sensing data and even unique in helping to identify surface materials and features such as rock types, soil moisture, geothermal anomalies etc. The ability to record variations in infrared radiation has advantage in extending our observation of many types of phenomena in which minor temperature variations may be significant in understanding our environment. Thermal remote sensing reserves immense potential for various applications. The following is a list of some of the areas in which thermal data is put to use:

- Identification of geological units and structures,

- Soil moisture studies,

- Hydrology,

- Coastal zones,

- Volcanology,

- Forest fires,

- Coal fires,

- Seismology,

- Environmental modelling,

- Meteorology,

- Medical sciences,

- Vetenary sciences,

- Intelligence / military applications,

- Heat loss from buildings.

LiDAR

LiDAR, or light detection and ranging, is a popular remote sensing method used for measuring the exact distance of an object on the earth's surface. Even though it was first used in the 1960s when laser scanners were mounted to aeroplanes, LiDAR didn't get the popularity it deserved until twenty years later. It was only during the 1980s after the introduction of GPS that it became a popular method for calculating accurate geospatial measurements.

Now that its scope has spread across numerous fields, we should know more about LiDAR mapping technology and how it works.

LiDAR Technology

According to the American Geoscience Institute, LiDAR uses a pulsed laser to calculate an object's variable distances from the earth surface. These light pulses — put together with the information collected by the airborne system — generate accurate 3D information about the earth surface and the target object.

There are three primary components of a LiDAR instrument — the scanner, laser and GPS receiver. Other elements that play a vital role in the data collection and analysis are the photo detector and optics. Most government and private organizations use helicopters, drones and airplanes for acquiring LiDAR data.

Types of LiDAR Systems

LiDAR systems are divided into two types based on its functionality — Airborne LiDAR and Terrestrial LiDAR.

Airborne LiDAR

Airborne LiDAR is installed on a helicopter or drone for collecting data. As soon as it's activated, Airborne LiDAR emits light towards the ground surface, which returns to the sensor immediately after hitting the object, giving an exact measurement of its distance. Airborne LiDAR is further divided into two types — Topological LiDAR and Bathymetric LiDAR.

Terrestrial LiDAR

Unlike Airborne, Terrestrial LiDAR systems are installed on moving vehicles or tripods on the earth surface for collecting accurate data points. These are quite common for observing highways, analysing infrastructure or even collecting point clouds from the inside and outside of buildings. Terrestrial LiDAR systems have two types — Mobile LiDAR and Static LiDAR.

How does LiDAR Work?

LiDAR follows a simple principle — throw laser light at an object on the earth surface and calculate the time it takes to return to the LiDAR source. Given the speed at which the light travels (approximately 186,000 miles per second), the process of measuring the exact distance through LiDAR appears to be incredibly fast. However, it's very technical. The formula that analysts use to arrive at the precise distance of the object is as follows:

- The distance of the object = (Speed of Light × Time of Flight)/2.

- LiDAR can be used to accomplish many developmental objectives, some of which are:

Oceanography

When the authorities want to know the exact depth of the ocean's surface to locate any object in the case of a maritime accident or for research purposes, they use LiDAR technology to accomplish their mission. Other than locating objects, LiDAR is also used for calculating phytoplankton fluorescence and biomass in the ocean surface, which otherwise is very challenging.

Digital Elevation or Terrain Model

Terrain elevations play a crucial role during the construction of roads, large buildings and bridges. LiDAR technology has x, y and z coordinates, which makes it incredibly easy to produce the 3D representation of elevations to ensure that concerned parties can draw necessary conclusions more easily.

Agriculture and Archaeology

Typical applications of LiDAR technology in the agriculture sector include analysis of yield rates, crop scouting and seed dispersions. Besides this, it is also used for campaign planning, mapping under the forest canopy, and more.

Apart from the applications mentioned above, LiDAR is used by geoscientists for unearthing geomorphology related secrets, as well as by military for carrying out various security operations near the national borders.

References

- Know-basics-of-remote-sensing, what-is-remote-sensing: grindgis.com, Retrieved 21 May, 2020

- Advantages-and-disadvantages-of-remote-sensin, remote-sensing: grindgis.com, Retrieved 22 June, 2020

- Elements-of-Remote-Sensing-Systems-1154: brainkart.com, Retrieved 23 July, 2020

- Tamop425/0027-DAI6, tamop425: tankonyvtar.hu, Retrieved 24 August, 2020

- What-is-lidar-technology-and-how-does-it-work: geospatialworld.net, Retrieved 25 January, 2020

2

Satellite Remote Sensing

Satellite remote sensing refers to the use of satellites for the estimation of geophysical parameters from the electromagnetic radiations emitted from Earth. IKONOS, SatNav, Envisat, SPOT, SCIAMACHY, etc. are some of its examples. This chapter discusses these types and related aspects of satellite remote sensing in detail.

Remote sensing satellites are equipped with sensors looking down to the earth. They are the "eyes in the sky" constantly observing the earth as they go round in predictable orbits. In satellite remote sensing of the earth, the sensors are looking through a layer of atmosphere separating the sensors from the Earth's surface being observed. Hence, it is essential to understand the effects of atmosphere on the electromagnetic radiation travelling from the Earth to the sensor through the atmosphere. The atmospheric constituents cause wavelength dependent absorption and scattering of radiation. These effects degrade the quality of images. Some of the atmospheric effects can be corrected before the images are subjected to further analysis and interpretation.

A consequence of atmospheric absorption is that certain wavelength bands in the electromagnetic spectrum are strongly absorbed and effectively blocked by the atmosphere. The wavelength regions in the electromagnetic spectrum usable for remote sensing are determined by their ability to penetrate atmosphere. These regions are known as the atmospheric transmission windows. Remote sensing systems are often designed to operate within one or more of the atmospheric windows. These windows exist in the microwave region, some wavelength bands in the infrared, the entire visible region and part of the near ultraviolet regions. Although the atmosphere is practically transparent to x-rays and gamma rays, these radiations are not normally used in remote sensing of the earth.

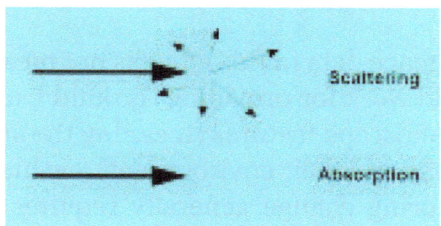

Technology of Satellite Remote Sensing

The fundamental principles of remote sensing derive from the characteristics and interactions of electromagnetic radiation (EMR) as it propagates from source to sensor. The principles relate to the following: 1) the source of energy and the type and amount of energy it provides; 2) the absorption and scattering effects of the atmosphere on EMR; 3) the mechanisms of EMR interaction with Earth surface features; and 4) the nature of sensor response as determined by the type of sensor.

Most satellite sensors detect EMR electronically as a continuous stream of digital data. The data are transmitted to ground reception stations, processed to create defined data products, and made available for sale to users on a variety of digital data media. Once purchased, the digital image data are readily amenable to quantitative analysis using computer-implemented digital image processing techniques. Some of these techniques (such as data error compensations, atmospheric corrections, calibration, and map registration) essentially involve pre-processing the data for subsequent interpretation and analysis. Another group of techniques is designed to selectively enhance the digital data and produce hard-copy image formats for interpreters to study. For these images, some of the principles and techniques of airphoto interpretation can be applied to manual analysis of the image information content. A third major group of digital processing techniques involves information extraction through the implementation of a wide range of simple to complex mathematical and statistical operations on the numerical data values in the image. The results of these operations provide output such as derived information variables (that might relate to terrain brightness or vegetation condition), categorized land and water features, or images showing changes over time.

A discussion of remote sensing technology would not be complete without mention of geographic information systems (GIS). Satellite remote sensing represents a technology for synoptic acquisition of spatial data and the extraction of scene-specific information. GIS provides a computer-implemented spatially oriented database for evaluating the information in conjunction with other spatially formatted data and information that may be acquired from remote sensor data, maps, surveys, and other sources of spatially referenced information. GIS technology should aid human dimensions studies of global change by enabling the integration and joint analysis of human science data and natural science data.

The emphasis on satellite-image data in this guide is not meant to diminish the roles of aerial photography and field work for providing "ground truth" data. Any comprehensive program of mapping landscape features (meaning the spatial manifestation of the relationship between humans and their environment), evaluating and quantifying their characteristics, and monitoring change generally requires supporting ground truth data to develop and verify the use of satellite-image data. Developing the intended use of satellite sensor data refers to establishing the qualitative associations or quantitative relationships one wants to implement; in other words, determining the capability to accomplish an objective using a particular type of data. Verification refers to assessment of performance and refinement of the capability. These activities enable establishing the link between satellite image data and the desired landscape information (both natural and human dimensions attributes).

The type of supporting ground truth data employed in specific studies varies relative to the scale of the primary data being used. When working with coarse-resolution continental- and global-scale satellite image data (such as Advanced Very High Resolution Radiometer data of 1-km spatial resolution), high-resolution satellite data (such as Landsat or Systeme Probatoire d'Observation de la Terra (SPOT)) may serve as supporting ground-truth data. On the other hand, scientists studying the human dimensions of global change with high-resolution satellite images may need to incorporate interpretations of airphotos and information acquired in the field to corroborate their intended use of the satellite data and validate their results.

Satellite Imagery

Satellite imagery (also Earth observation imagery or spaceborne photography) are images of Earth or other planets collected by imaging satellites operated by governments and businesses around the world. Satellite imaging companies sell images by licensing them to governments and businesses such as Apple Maps and Google Maps.

Uses of Satellite Imagery

Satellite photography can be used to produce composite images of an entire hemisphere.

Map a small area of the Earth.

Satellite images have many applications in meteorology, oceanography, fishing, agriculture, biodiversity conservation, forestry, landscape, geology, cartography, regional planning, education, intelligence and warfare. Less mainstream uses include anomaly hunting, a criticized investigation technique involving the search of satellite images for unexplained phenomena. Images can be in visible colors and in other spectra. There are also elevation maps, usually made by radar images. Interpretation and analysis of satellite imagery is conducted using specialized remote sensing software.

Data Characteristics

There are four types of resolution when discussing satellite imagery in remote sensing: spatial, spectral, temporal, and radiometric. Campbell defines these as follows:

- Spatial resolution is defined as the pixel size of an image representing the size of the surface area (i.e. M^2) being measured on the ground, determined by the sensors' instantaneous field of view (ifov).

- Spectral resolution is defined by the wavelength interval size (discrete segment of the electromagnetic spectrum) and number of intervals that the sensor is measuring.

- Temporal resolution is defined by the amount of time (e.g. days) that passes between imagery collection periods for a given surface location.

- Radiometric resolution is defined as the ability of an imaging system to record many levels of brightness (contrast for example) and to the effective bit-depth of the sensor (number of grayscale levels) and is typically expressed as 8-bit (0–255), 11-bit (0–2047), 12-bit (0–4095) or 16-bit (0–65,535).

- Geometric resolution refers to the satellite sensor's ability to effectively image a portion of the Earth's surface in a single pixel and is typically expressed in terms of Ground sample distance, or GSD. GSD is a term containing the overall optical and systemic noise sources and is useful for comparing how well one sensor can "see" an object on the ground within a single pixel. For example, the GSD of Landsat is ≈30m, which means the smallest unit that maps to a single pixel within an image is ≈30m x 30m. The latest commercial satellite (GeoEye 1) has a GSD of 0.41 m. This compares to a 0.3 m resolution obtained by some early military film based Reconnaissance satellite such as Corona.

The resolution of satellite images varies depending on the instrument used and the altitude of the satellite's orbit. For example, the Landsat archive offers repeated imagery at 30 meter resolution for the planet, but most of it has not been processed from the raw data. Landsat 7 has an average return period of 16 days. For many smaller areas, images with resolution as high as 41 cm can be available.

Satellite imagery is sometimes supplemented with aerial photography, which has higher resolution, but is more expensive per square meter. Satellite imagery can be combined with vector or raster data in a GIS provided that the imagery has been spatially rectified so that it will properly align with other data sets.

Imaging Satellites

Public Domain

Satellite imaging of the Earth surface is of sufficient public utility that many countries maintain satellite imaging programs.

Landsat

Landsat is the oldest continuous Earth observing satellite imaging program. Optical Landsat imagery has been collected at 30 m resolution since the early 1980s. Beginning with Landsat 5, thermal infrared imagery was also collected (at coarser spatial resolution than the optical data). The Landsat 7 and Landsat 8 satellites are currently in orbit. Landsat 9 is planned.

MODIS

MODIS has collected near-daily satellite imagery of the earth in 36 spectral bands since 2000. MODIS is on board the NASA Terra and Aqua satellites.

Sentinel

The ESA is currently developing the Sentinel constellation of satellites. Currently, 7 missions are planned, each for a different application. Sentinel-1 (SAR imaging), Sentinel-2 (decameter optical imaging for land surfaces), and Sentinel-3 (hectometer optical and thermal imaging for land and water) have already been launched.

ASTER

The Advanced Spaceborne Thermal Emission and Reflection Radiometer (ASTER) is an imaging instrument onboard Terra, the flagship satellite of NASA's Earth Observing

System (EOS) launched in December 1999. ASTER is a cooperative effort between NASA, Japan's Ministry of Economy, Trade and Industry (METI), and Japan Space Systems (J-spacesystems). ASTER data is used to create detailed maps of land surface temperature, reflectance, and elevation. The coordinated system of EOS satellites, including Terra, is a major component of NASA's Science Mission Directorate and the Earth Science Division. The goal of NASA Earth Science is to develop a scientific understanding of the Earth as an integrated system, its response to change, and to better predict variability and trends in climate, weather, and natural hazards.

- Land surface climatology: Investigation of land surface parameters, surface temperature, etc., to understand land-surface interaction and energy and moisture fluxes.

- Vegetation and ecosystem dynamics: Investigations of vegetation and soil distribution and their changes to estimate biological productivity, understand land-atmosphere interactions, and detect ecosystem change.

- Volcano monitoring: Monitoring of eruptions and precursor events, such as gas emissions, eruption plumes, development of lava lakes, eruptive history and eruptive potential.

- Hazard monitoring: Observation of the extent and effects of wildfires, flooding, coastal erosion, earthquake damage, and tsunami damage.

- Hydrology: Understanding global energy and hydrologic processes and their relationship to global change; included is evapotranspiration from plants.

- Geology and soils: The detailed composition and geomorphologic mapping of surface soils and bedrocks to study land surface processes and earth's history.

- Land surface and land cover change: Monitoring desertification, deforestation, and urbanization; providing data for conservation managers to monitor protected areas, national parks, and wilderness areas.

Meteosat

Model of a first generation Meteosat geostationary satellite.

The Meteosat-2 geostationary weather satellite began operationally to supply imagery data on 16 August 1981. Eumetsat has operated the Meteosats since 1987.

- The Meteosat visible and infrared imager (MVIRI), three-channel imager: visible, infrared and water vapour; It operates on the first generation Meteosat, Meteosat-7 being still active.

- The 12-channel Spinning Enhanced Visible and Infrared Imager (SEVIRI) includes similar channels to those used by MVIRI, providing continuity in climate data over three decades; Meteosat Second Generation (MSG).

- The Flexible Combined Imager (FCI) on Meteosat Third Generation (MTG) will also include similar channels, meaning that all three generations will have provided over 60 years of climate data.

Private Domain

Several satellites are built and maintained by private companies. These include:

GeoEye

GeoEye's GeoEye-1 satellite was launched on September 6, 2008. The GeoEye-1 satellite has the high resolution imaging system and is able to collect images with a ground resolution of 0.41 meters (16 inches) in the panchromatic or black and white mode. It collects multispectral or color imagery at 1.65-meter resolution or about 64 inches.

DigitalGlobe

DigitalGlobe's WorldView-2 satellite provides high resolution commercial satellite imagery with 0.46 m spatial resolution (panchromatic only). The 0.46 meters resolution of WorldView-2's panchromatic images allows the satellite to distinguish between objects on the ground that are at least 46 cm apart. Similarly DigitalGlobe's QuickBird satellite provides 0.6 meter resolution (at NADIR) panchromatic images.

DigitalGlobe's WorldView-3 satellite provides high resolution commercial satellite imagery with 0.31 m spatial resolution. WVIII also carries a short wave infrared sensor and an atmospheric sensor.

SPOT Image

The 3 SPOT satellites in orbit (Spot 5, 6, 7) provide very high resolution images – 1.5 m for Panchromatic channel, 6m for Multi-spectral (R,G,B,NIR). Spot Image also distributes multiresolution data from other optical satellites, in particular from Formosat-2 (Taiwan) and Kompsat-2 (South Korea) and from radar satellites (TerraSar-X, ERS, Envisat, Radarsat). Spot Image is also the exclusive distributor of data from the high resolution Pleiades satellites with a resolution of 0.50 meter or about 20 inches. The launches occurred in 2011 and 2012, respectively. The company also offers infrastructures for receiving and processing, as well as added value options.

SPOT image of Bratislava.

BlackBridge

BlackBridge, previously known as RapidEye, operates a constellation of five satellites, launched in August 2008, the RapidEye constellation contains identical multispectral sensors which are equally calibrated. Therefore, an image from one satellite will be equivalent to an image from any of the other four, allowing for a large amount of imagery to be collected (4 million km² per day), and daily revisit to an area. Each travel on the same orbital plane at 630 km, and deliver images in 5 meter pixel size. RapidEye satellite imagery is especially suited for agricultural, environmental, cartographic and disaster management applications. The company not only offers their imagery, but consults their customers to create services and solutions based on analysis of this imagery.

ImageSat International

Earth Resource Observation Satellites, better known as "EROS" satellites, are lightweight, low earth orbiting, high-resolution satellites designed for fast maneuvering between imaging targets. In the commercial high-resolution satellite market, EROS is the smallest very high resolution satellite; it is very agile and thus enables very high performances. The satellites are deployed in a circular sun-synchronous near polar orbit at an altitude of 510 km (+/- 40 km). EROS satellites imagery applications are primarily for intelligence, homeland security and national development purposes but also employed in a wide range of civilian applications, including: mapping, border control, infrastructure planning, agricultural monitoring, environmental monitoring, disaster response, training and simulations, etc.

EROS A – a high resolution satellite with 1.9–1.2m resolution panchromatic was launched on December 5, 2000.

EROS B – the second generation of Very High Resolution satellites with 70 cm resolution panchromatic, was launched on April 25, 2006.

China Siwei

GaoJing-1/SuperView-1 (01, 02, 03, 04) is a commercial constellation of Chinese remote sensing satellites controlled by China Siwei Surveying and Mapping Technology Co. Ltd. The four satellites operate from an altitude of 530km and are phased 90° from each other on the same orbit, providing 0.5m panchromatic resolution and 2m multispectral resolution on a swath of 12 km.

Imagery Analysis using Artificial Intelligence

Advancements in artificial intelligence have made autonomous, large scale analysis of imagery possible. AI has been taught to process Satellite Imagery with a small degree of error. Studies have used AI to differentiate between different forest types and AI can tell the difference between certain soil and vegetation types. Researchers are using AI to monitor Satellite Imagery for vineyard and grape health as well as having AI estimate wheat harvest size. Projects like SpaceKnow uses AI to conduct case studies in near real-time of deforestation due to wildfires in California and manufacturing activity in China.

As the technology advances, clearer imagery and faster neural networks has allowed for the study of Above Ground Biomass (AGB). This ABG index can describe the size and density of vegetation which scientists use to estimate carbon output and footprints in certain areas. Scientists are eager to apply this data to the study of global warming and climate change. Researchers are developing AI that can monitor refugee movements in war-torn countries, monitor deforestation in the Amazon rain-forest, and show algae blooms in places like the Gulf of Mexico and the Red Sea. Upcoming studies of contaminated surface water and chemical runoff from Fracking are also being planned.

Disadvantages

Because the total area of the land on Earth is so large and because resolution is relatively high, satellite databases are huge and image processing (creating useful images from the raw data) is time-consuming. Preprocessing, such as image destriping is often required. Depending on the sensor used, weather conditions can affect image quality: for example, it is difficult to obtain images for areas of frequent cloud cover such as mountain-tops. For such reasons, publicly available satellite image datasets are typically processed for visual or scientific commercial use by third parties.

Commercial satellite companies do not place their imagery into the public domain and do not sell their imagery; instead, one must be licensed to use their imagery. Thus, the ability to legally make derivative products from commercial satellite imagery is minimized.

Multiple Satellite Imaging

Multiple satellite imaging is the process of using multiple satellites to gather more information than a single satellite so that a better estimate of the desired source is possible. So something that cannot be seen with one telescope might be visible with two or more telescopes.

Space Interferometry Mission conceptual picture.

Interferometry is the process of combining waves in such a way that they constructively interfere. When two or more independent sources detect a signal at the same given frequency those signals can be combined and the result is better than each one individually.

The NASA Origins Program was created in the 1990s to ultimately search for the origin of the universe. The theory that the Origins Program is based on is: since light travels at a constant speed until it is absorbed by something; there is still light that was part of the first light ever created traveling about the universe and ultimately some of that light is coming in the general direction of Earth. So a satellite system capable of collecting light from the beginning of the universe would be able to tell us more about where we came from.

There is also the constant search for life in other worlds. A satellite system using the interferometric technologies mentioned above would be able to have a much higher resolution than any of the current deep space imaging systems. A space systems also reduces the amount of interference due to lack of an atmosphere.

Intelligence Satellite

intelligence satellite (commonly, although unofficially, referred to as a spy satellite) is an Earth observation satellite or communications satellite deployed for military or intelligence applications.

The first generation type (i.e., Corona and Zenit) took photographs, then ejected canisters of photographic film which would descend back down into Earth's atmosphere. Corona capsules were retrieved in mid-air as they floated down on parachutes. Later, spacecraft had digital imaging systems and downloaded the images via encrypted radio links.

Most information available is on programs that existed up to 1972, as this information has been declassified due to its age. Some information about programs prior to that time is still classified, and a small amount of information is available on subsequent missions.

A few up-to-date reconnaissance satellite images have been declassified on occasion, or leaked.

Types

There are several major types of reconnaissance satellite:

- Missile early warning: Provides warning of an attack by detecting ballistic missile launches. Earliest known are Missile Defense Alarm System.

- Nuclear explosion detection: Identifies and characterizes nuclear explosions in space. Vela (satellite) is the earliest known.

- Photo surveillance: Provides imaging of earth from space. Images can be a survey or close-look telephoto. Corona (satellite) is the earliest known. Spectral imaging is commonplace.

- Electronic reconnaissance: Signals intelligence, intercepts stray radio waves. Samos-F is the earliest known.

- Radar imaging: Most space-based radars use synthetic aperture radar. Can be used at night or through cloud cover. Earliest known are the Soviet US-A series.

Missions

Examples of reconnaissance satellite missions:

- High resolution photography (IMINT).

- Measurement and Signature Intelligence (MASINT).

- Communications eavesdropping (SIGINT).

- Covert communications.

- Monitoring of nuclear test ban compliance.

- Detection of missile launches.

On 17 February 2014, a Russian Kosmos-1220 originally launched in 1980 and used for naval missile targeting until 1982, made an uncontrolled atmospheric entry.

Benefits

Reconnaissance satellites have been used to enforce human rights, through the Satellite Sentinel Project, which monitors atrocities in Sudan and South Sudan.

During his 1980 State of the Union Address, President Jimmy Carter explained how all of humanity benefited from the presence of American spy satellites:

> Photo-reconnaissance satellites, for example, are enormously important in stabilizing world affairs and thereby make a significant contribution to the security of all nations.

Additionally, companies such as GeoEye and DigitalGlobe have provided commercial satellite imagery in support of natural disaster response and humanitarian missions.

IKONOS

IKONOS was a commercial Earth observation satellite, and was the first to collect publicly available high-resolution imagery at 1- and 4-meter resolution. It collected multispectral (MS) and panchromatic (PAN) imagery. The capability to observe Earth via space-based telescope has been called "one of the most significant developments in the history of the space age", and IKONOS brought imagery rivaling that of military spy satellites to the commercial market. IKONOS imagery began being sold on 1 January 2000, and the spacecraft was retired in 2015.

Specifications

Spacecraft

IKONOS was a three-axis stabilized spacecraft designed by Lockheed Martin Space Systems. The design later became known as the LM-900 satellite bus and was optimized to carry remote sensing payloads. Four reaction wheels stabilized the spacecraft's attitude, which was measured by two star trackers and a sun sensor. Orbital position information was provided by a GPS receiver. The spacecraft body was a hexagonal design of 1.83 by 1.57 meters (6.0 by 5.2 ft) and 817 kilograms (1,800 lb), with 1.5 kilowatts of power provided by three solar panels. Its design life was seven years. *IKONOS* operated in a Sun-synchronous, near-polar, circular orbit at approximately 680 km (423 mi).

Optical Sensor Assembly

IKONOS's primary instrument was the Optical Sensor Assembly (OSA), designed and built by Kodak. It had a primary mirror aperture of 70 cm (28 in), and a folded optical focal length of 10 m (394 in) using 5 mirrors. The main mirror featured a honeycomb design to reduce mass. The detectors at the focal plane included a panchromatic sensor with 13,500 pixels cross-track, and four multispectral sensors (blue, green, red, and near-infrared) each with 3,375 pixels along-track. Its nadir image swath was 11.3 km (7 mi). Total instrument mass was 171 kg (377 lb) and it consumed 350 watts.

SatNav

A satellite navigation or SatNav system is a system that uses satellites to provide autonomous geo-spatial positioning. It allows small electronic receivers to determine their location (longitude, latitude, and altitude/elevation) to high precision (within a few centimeters to metres) using time signals transmitted along a line of sight by radio from satellites. The system can be used for providing position, navigation or for tracking the position of something fitted with a receiver (satellite tracking). The signals also allow the electronic receiver to calculate the current local time to high precision, which allows time synchronisation. Satnav systems operate independently of any telephonic or internet reception, though these technologies can enhance the usefulness of the positioning information generated.

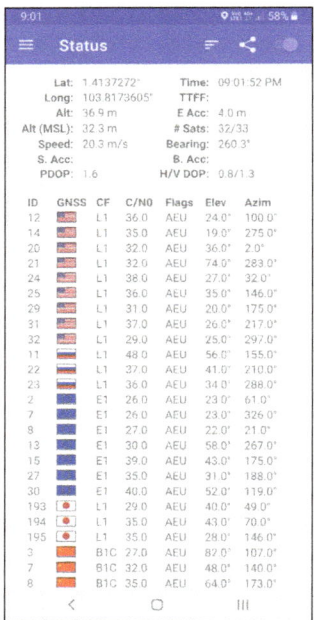

GPSTest showing the available GNSS in 2019. Since the 2010s, satellite navigation is widely available on civilian devices.

A satellite navigation system with global coverage may be termed a global navigation satellite system (GNSS). As of October 2018, the United States' Global Positioning System (GPS) and Russia's Global Navigation Satellite System (GLONASS) are fully operational GNSSs, with China's BeiDou Navigation Satellite System (BDS) and the

European Union's Galileo scheduled to be fully operational by 2020. Japan's Quasi-Zenith Satellite System (QZSS) is a GPS satellite-based augmentation system to enhance GPS's accuracy, with satellite navigation independent of GPS scheduled for 2023.

Global coverage for each system is generally achieved by a satellite constellation of 18–30 medium Earth orbit (MEO) satellites spread between several orbital planes. The actual systems vary, but use orbital inclinations of >50° and orbital periods of roughly twelve hours (at an altitude of about 20,000 kilometres or 12,000 miles).

Ground based radio navigation has long been practiced. The DECCA, LORAN, GEE and Omega systems used terrestrial longwave radio transmitters which broadcast a radio pulse from a known "master" location, followed by a pulse repeated from a number of "slave" stations. The delay between the reception of the master signal and the slave signals allowed the receiver to deduce the distance to each of the slaves, providing a fix.

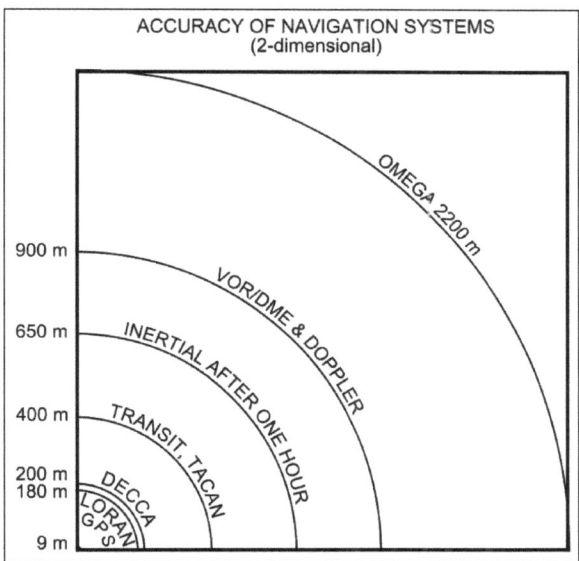

The first satellite navigation system was Transit, a system deployed by the US military in the 1960s. Transit's operation was based on the Doppler effect: the satellites travelled on well-known paths and broadcast their signals on a well-known radio frequency. The received frequency will differ slightly from the broadcast frequency because of the movement of the satellite with respect to the receiver. By monitoring this frequency shift over a short time interval, the receiver can determine its location to one side or the other of the satellite, and several such measurements combined with a precise knowledge of the satellite's orbit can fix a particular position. Satellite orbital position errors are induced by variations in the gravity field and radar refraction, among others. These were resolved by a team led by Harold L Jury of Pan Am Aerospace Division in Florida from 1970-1973. Using real-time data assimilation and recursive estimation, the systematic and residual errors were narrowed down to a manageable level to permit accurate navigation.

Part of an orbiting satellite's broadcast included its precise orbital data. In order to ensure accuracy, the US Naval Observatory (USNO) continuously observed the precise orbits of these satellites. As a satellite's orbit deviated, the USNO would send the updated information to the satellite. Subsequent broadcasts from an updated satellite would contain its most recent ephemeris.

Modern systems are more direct. The satellite broadcasts a signal that contains orbital data (from which the position of the satellite can be calculated) and the precise time the signal was transmitted. Orbital data include a rough almanac for all satellites to aid in finding them, and a precise ephemeris for this satellite. The orbital ephemeris is transmitted in a data message that is superimposed on a code that serves as a timing reference. The satellite uses an atomic clock to maintain synchronization of all the satellites in the constellation. The receiver compares the time of broadcast encoded in the transmission of three (at sea level) or four different satellites, thereby measuring the time-of-flight to each satellite. Several such measurements can be made at the same time to different satellites, allowing a continual fix to be generated in real time using an adapted version of trilateration.

Each distance measurement, regardless of the system being used, places the receiver on a spherical shell at the measured distance from the broadcaster. By taking several such measurements and then looking for a point where they meet, a fix is generated. However, in the case of fast-moving receivers, the position of the signal moves as signals are received from several satellites. In addition, the radio signals slow slightly as they pass through the ionosphere, and this slowing varies with the receiver's angle to the satellite, because that changes the distance through the ionosphere. The basic computation thus attempts to find the shortest directed line tangent to four oblate spherical shells centred on four satellites. Satellite navigation receivers reduce errors by using combinations of signals from multiple satellites and multiple correlators, and then using techniques such as Kalman filtering to combine the noisy, partial, and constantly changing data into a single estimate for position, time, and velocity.

Applications

Automotive navigation system.

The original motivation for satellite navigation was for military applications. Satellite navigation allows precision in the delivery of weapons to targets, greatly increasing their lethality whilst reducing inadvertent casualties from mis-directed weapons. Satellite navigation also allows forces to be directed and to locate themselves more easily, reducing the fog of war.

Now a global navigation satellite system, such as Galileo, is used to determine users location and the location of other people or objects at any given moment. The range of application of the satellite in the future is enormous, including both the public and private sectors across numerous market segments such as science, transport, agriculture etc.

The ability to supply satellite navigation signals is also the ability to deny their availability. The operator of a satellite navigation system potentially has the ability to degrade or eliminate satellite navigation services over any territory it desires.

Global Navigation Satellite Systems

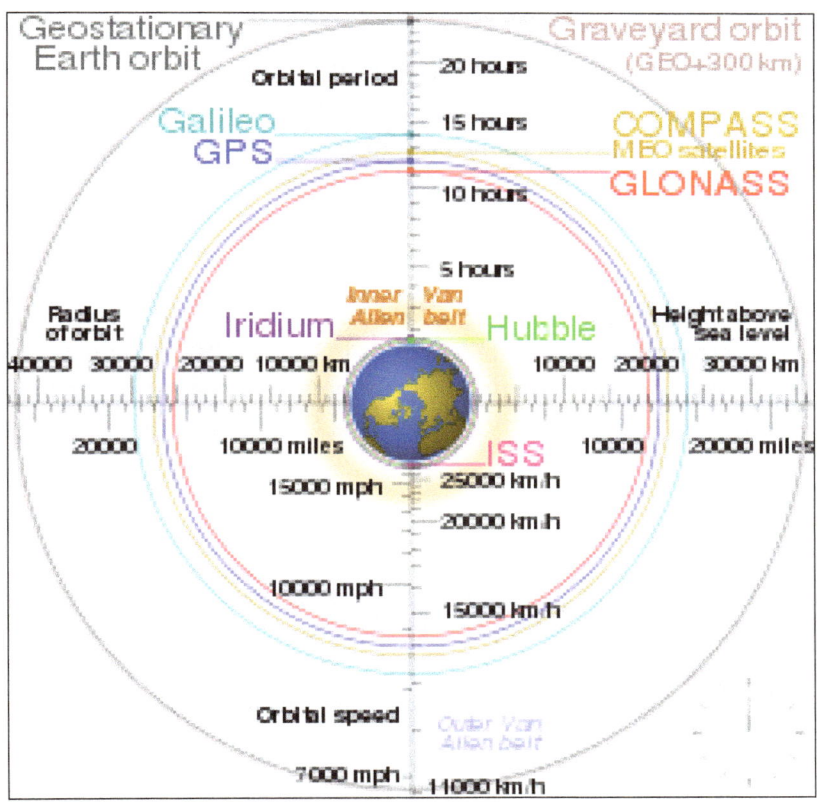

Comparison of geostationary, GPS, GLONASS, Galileo, Compass (MEO), International Space Station, Hubble Space Telescope, Iridium constellation and graveyard orbits, with the Van Allen radiation belts and the Earth to scale. The Moon's orbit is around 9 times as large as geostationary orbit.

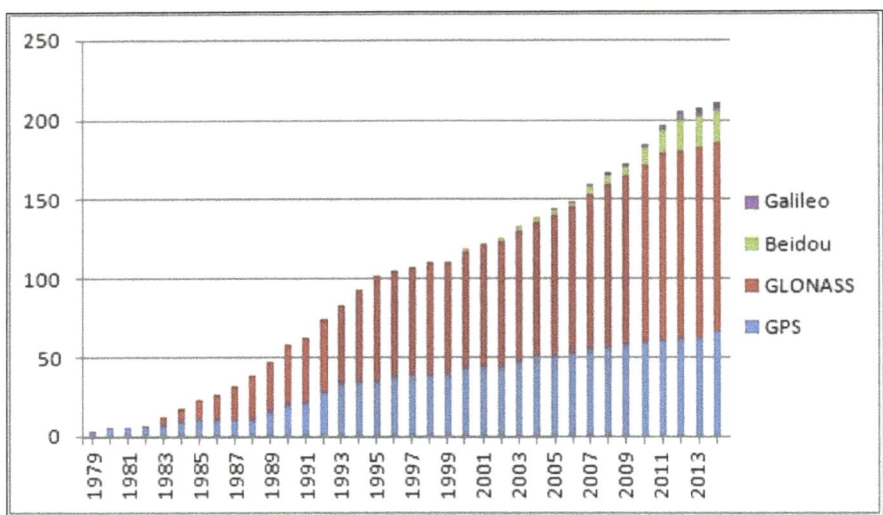

Launched GNSS satellites 1978 to 2014.

GPS

The United States' Global Positioning System (GPS) consists of up to 32 medium Earth orbit satellites in six different orbital planes, with the exact number of satellites varying as older satellites are retired and replaced. Operational since 1978 and globally available since 1994, GPS is the world's most utilized satellite navigation system.

GLONASS

The formerly Soviet, and now Russian, *Global'naya Navigatsionnaya Sputnikovaya Sistema,* (GLObal NAvigation Satellite System or GLONASS), is a space-based satellite navigation system that provides a civilian radionavigation-satellite service and is also used by the Russian Aerospace Defence Forces. GLONASS has full global coverage with 24 satellites.

Galileo

The European Union and European Space Agency agreed in March 2002 to introduce their own alternative to GPS, called the Galileo positioning system. Galileo became operational on 15 December 2016 (global Early Operational Capability (EOC)) At an estimated cost of €3 billion, the system of 30 MEO satellites was originally scheduled to be operational in 2010. The original year to become operational was 2014. The first experimental satellite was launched on 28 December 2005. Galileo is expected to be compatible with the modernized GPS system. The receivers will be able to combine the signals from both Galileo and GPS satellites to greatly increase the accuracy. Galileo is expected to be in full service in 2020 and at a substantially higher cost. The main modulation used in Galileo Open Service signal is the Composite Binary Offset Carrier (CBOC) modulation.

BeiDou

Beidou started as the now-decommissioned Beidou-1, an Asia-Pacific local network on the geostationary orbits. China has indicated their plan to complete the entire second generation Beidou Navigation Satellite System (BDS or BeiDou-2, formerly known as COMPASS), by expanding current regional (Asia-Pacific) service into global coverage by 2020. This BeiDou-3 system is proposed to consist of 30 MEO satellites and five geostationary satellites (IGSO). A 16-satellite regional version (covering Asia and Pacific area) was completed by December 2012. Global service was completed by December 2018.

Augmentation

GNSS augmentation is a method of improving a navigation system's attributes, such as accuracy, reliability, and availability, through the integration of external information into the calculation process, for example, the Wide Area Augmentation System, the European Geostationary Navigation Overlay Service, the Multi-functional Satellite Augmentation System, Differential GPS, GPS-aided GEO augmented navigation (GAGAN) and inertial navigation systems.

DORIS

Doppler Orbitography and Radio-positioning Integrated by Satellite (DORIS) is a French precision navigation system. Unlike other GNSS systems, it is based on static emitting stations around the world, the receivers being on satellites, in order to precisely determine their orbital position. The system may be used also for mobile receivers on land with more limited usage and coverage. Used with traditional GNSS systems, it pushes the accuracy of positions to centimetric precision (and to millimetric precision for altimetric application and also allows monitoring very tiny seasonal changes of Earth rotation and deformations), in order to build a much more precise geodesic reference system.

Low Earth Orbit Satellite Phone Networks

The two current operational low Earth orbit satellite phone networks are able to track transceiver units with accuracy of a few kilometers using doppler shift calculations from the satellite. The coordinates are sent back to the transceiver unit where they can be read using AT commands or a graphical user interface. This can also be used by the gateway to enforce restrictions on geographically bound calling plans.

Envisat

Envisat ("Environmental Satellite") is a large inactive Earth-observing satellite which is still in orbit. Operated by the European Space Agency (ESA), it was the world's largest civilian Earth observation satellite.

It was launched on 1 March 2002 aboard an Ariane 5 from the Guyana Space Centre in Kourou, French Guiana, into a Sun synchronous polar orbit at an altitude of 790 ± 10 km. It orbits the Earth in about 101 minutes, with a repeat cycle of 35 days. After losing contact with the satellite on 8 April 2012, ESA formally announced the end of Envisat's mission on 9 May 2012.

Envisat cost 2.3 billion Euro (including 300 million Euro for 5 years of operations) to develop and launch. The mission has been replaced by the Sentinel series of satellites. The first of these, Sentinel 1, has taken over the radar duties of Envisat since its launch in 2014.

Mission

Envisat was launched as an Earth observation satellite. Its objective was to service the continuity of European Remote-Sensing Satellite missions, providing additional observational parameters to improve environmental studies.

In working towards the global and regional objectives of the mission, numerous scientific disciplines currently use the data acquired from the different sensors on the satellite to study such things as atmospheric chemistry, ozone depletion, biological oceanography, ocean temperature and colour, wind waves, hydrology (humidity, floods), agriculture and arboriculture, natural hazards, digital elevation modelling (using interferometry), monitoring of maritime traffic, atmospheric dispersion modelling (pollution), cartography and study of snow and ice.

Specifications

Dimensions

26 m (85 ft) × 10 m (33 ft) × 5 m (16 ft).

Mass

8,211 kg (18,102 lb), including 319 kg (703 lb) of fuel and a 2,118 kg (4,669 lb) instrument payload.

Power

Solar array with a total load of 3560 W.

Instruments

Envisat carries an array of nine Earth-observation instruments that gathered information about the Earth (land, water, ice, and atmosphere) using a variety of measurement principles. A tenth instrument, DORIS, provided guidance and control. Several of the instruments were advanced versions of instruments that were flown on the earlier ERS 1 and ERS 2 missions and other satellites.

MWR

MWR (Microwave Radiometer) was designed for measuring water vapour in the atmosphere.

AATSR

AATSR (Advanced Along Track Scanning Radiometer) can measure the sea surface temperature in the visible and infrared spectra. Because of its wide angle lens it is possible to make very precise measurements of atmospheric effects on how emissions from the Earth's surface propagate.

AATSR is the successor of ATSR1 and ATSR2, payloads of ERS 1 and ERS 2. AATSR can measure Earth's surface temperature to a precision of 0.3 K (0.54 °F), for climate research. Among the secondary objectives of AATSR is the observation of environmental parameters such as water content, biomass, and vegetal health and growth.

MIPAS

MIPAS (Michelson Interferometer for Passive Atmospheric Sounding) is a Fourier transforming infrared spectrometer which provides pressure and temperature profiles, and profiles of trace gases nitrogen dioxide (NO_2), nitrous oxide (N_2O), methane (CH_4), nitric acid (HNO_3), ozone (O_3), and water (H_2O) in the stratosphere. The instrument functions with high spectral resolution in an extended spectral band, which allows coverage across the Earth in all seasons and at equal quality night and day. MIPAS has a vertical resolution of 3 to 5 km (2 to 3 mi) depending on altitude (the larger at the level of the upper stratosphere).

MERIS

MERIS (MEdium Resolution Imaging Spectrometer) measures the reflectance of the Earth (surface and atmosphere) in the solar spectral range (390 to 1040 nm) and transmits 15 spectral bands back to the ground segment. MERIS was built at the Cannes Mandelieu Space Center.

SCIAMACHY

SCIAMACHY (SCanning Imaging Absorption spectroMeter for Atmospheric CHartographY) compares light coming from the sun to light reflected by the Earth, which provides information on the atmosphere through which the Earth-reflected light has passed.

SCIAMACHY is an image spectrometer with the principal objective of mapping the concentration of trace gases and aerosols in the troposphere and stratosphere. Rays of sunlight that are reflected transmitted, backscattered and reflected by the atmosphere are captured at a high spectral resolution (0.2 to 0.5 nm) for wavelengths between 240 and 1700 nm, and in certain spectra between 2,000 and 2,400 nm. Its high spectral resolution over a wide range of wavelengths can detect many trace gases even in tiny concentrations. The wavelengths captured also allow effective detection of aerosols and clouds. SCIAMACHY uses 3 different targeting modes: to the nadir (against the sun), to the limbus (through the atmospheric corona), and during solar or lunar eclipses. SCIAMACHY was built by Netherlands and Germany at TNO/TPD, SRON and Airbus Defence and Space Netherlands.

RA-2

RA-2 (Radar Altimeter 2) is a dual-frequency Nadir pointing Radar operating in the K_u band and S bands, it is used to define ocean topography, map/monitor sea ice and measure land heights.

ASAR

ASAR (Advanced Synthetic Aperture Radar) operates in the C band in a wide variety of modes. It can detect changes in surface heights with sub-millimeter precision. It served as a data link for ERS 1 and ERS 2, providing numerous functions such as observations of different polarities of light or combining different polarities, angles of incidence and spatial resolutions.

Mode	Id	Polarisation	Incidence	Resolution	Swath
Alternating polarisation	AP	HH/VV, HH/HV, VV/VH	$15 - 45°$	30 – 150 m	58 – 110 km
Image	IM	HH, VV	$15 - 45°$	30 – 150 m	58 – 110 km
Wave	WV	HH, VV		400 m	5 km × 5 km
Suivi global (ScanSAR)	GM	HH, VV		1000 m	405 km
Wide Swath (ScanSAR)	WS	HH, VV		150 m	405 km

These different types of raw data can be given several levels of treatment (suffixed to the ID of the acquisition mode: IMP, APS, and so on):

- RAW (raw data, or "Level 0"), which contains all the information necessary to create images.

- S (complex data, "Single Look Complex"), images in complex numeric form, the real and imaginary parts of the output of the compression algorithm.

- P (precision image), amplified image with constant pixel width (12.5 m for IMP).

- M (medium precision image), amplified radiometry image with a resolution greater than P.

- G (geocoded image), amplified image to which simple geographical transforms have been applied to show relief.

Data capture in WV mode is unusual in that they constitute a series of 5 km × 5 km spaced at 100 km.

DORIS

DORIS (Doppler Orbitography and Radiopositioning Integrated by Satellite) determines the satellite's orbit to within 10 cm (4 in).

GOMOS

GOMOS (Global Ozone Monitoring by Occultation of Stars) looks at stars as they descend through the Earth's atmosphere and change color, allowing measurement of gases such as ozone (O_3), including their vertical distribution.

GOMOS uses the principle of occultation. Its sensors detect light from a star traversing the Earth's atmosphere and measures the depletion of that light by trace gases nitrogen dioxide (NO_2), nitrogen trioxide, (NO_3), OClO), ozone (O_3) and aerosols present between about 20 to 80 km (12 to 50 mi) altitude. It has a resolution of 3 km (1.9 mi).

Loss of Contact

ESA announced on 12 April 2012 that they lost contact with Envisat on Sunday, 8 April 2012, after 10 years of service, exceeding initially planned life span by 5 years. The spacecraft was still in a stable orbit, but attempts to contact it were unsuccessful. Ground-based radar and the French Pleiades Earth probe were used to image the silent Envisat and look for damage. ESA formally announced the end of Envisat's mission on 9 May 2012.

Envisat was launched in 2002 and it operated five years beyond its planned mission lifetime, delivering over a petabyte of data. ESA was expecting to turn off the spacecraft in 2014.

Space Safety

Envisat poses a hazard because of the risk of collisions with space debris. Given its orbit and its area-to-mass ratio, it will take about 150 years for the satellite to be gradually pulled into the Earth's atmosphere. Envisat is currently orbiting in an environment where two catalogued space debris objects can be expected to pass within about 200 m (660 ft) of it every year, which would likely trigger the need for a manoeuvre to avoid a possible collision. A collision between a satellite the size of Envisat and an object as small as 10 kg could produce a very large cloud of debris, initiating a self-sustaining chain-reaction of collisions and fragmentation with production of new debris, a phenomenon known as the Kessler Syndrome.

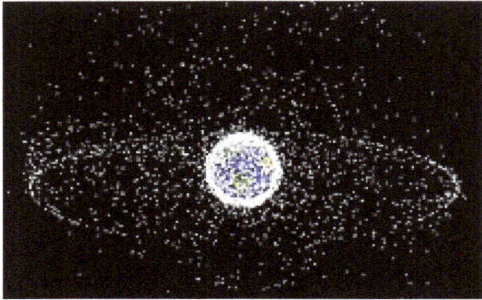

Space debris populations seen from outside geosynchronous orbit (GEO). Note the two primary debris fields, the ring of objects in GEO, and the cloud of objects in low Earth orbit (LEO).

Envisat is a candidate for a mission to remove it from orbit, called e.Deorbit. The spacecraft sent to bring down Envisat would itself need to have a mass of approximately 1.6 tonnes.

SPOT

SPOT is a commercial high-resolution optical imaging Earth observation satellite system operating from space. It is run by Spot Image, based in Toulouse, France. It was initiated by the CNES (*Centre national d'études spatiales* – the French space agency) in the 1970s and was developed in association with the SSTC (Belgian scientific, technical and cultural services) and the Swedish National Space Board (SNSB). It has been designed to improve the knowledge and management of the Earth by exploring the Earth's resources, detecting and forecasting phenomena involving climatology and oceanography, and monitoring human activities and natural phenomena. The SPOT system includes a series of satellites and ground control resources for satellite control and programming, image production, and distribution. Earlier satellites were launched using the European Space Agency's Ariane 2, 3, and 4 rockets,

Spot-5 Satellite.

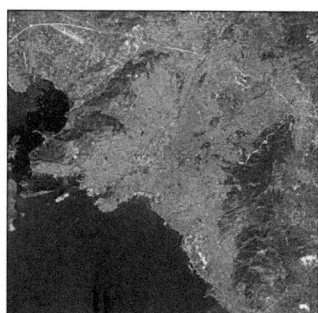

Athens as seen by the SPOT 5 satellite in 2002.

List of Satellites

SPOT Image is marketing the high-resolution images, which SPOT can take from every corner of the Earth.

- SPOT 1 launched February 22, 1986 with 10 panchromatic and 20 meter multi-spectral picture resolution capability. Withdrawn December 31, 1990.

- SPOT 2 launched January 22, 1990 and deorbited in July 2009.

- SPOT 3 launched September 26, 1993. Stopped functioning November 14, 1997.

- SPOT 4 launched March 24, 1998. Stopped functioning July, 2013.

- SPOT 5 launched May 4, 2002 with 2.5 m, 5 m and 10 m capability. Stopped functioning March 31, 2015.

- SPOT 6 launched September 9, 2012.

- SPOT 7 launched on June 30, 2014.

Orbit

The SPOT orbit is polar, circular, sun-synchronous, and phased. The inclination of the orbital plane combined with the rotation of the Earth around the polar axis allows the satellite to fly over any point on Earth within 26 days. The orbit has an altitude of 832 kilometers, an inclination of 98.7°, and completing $14 + 5/26$ revolutions per day.

Generations

SPOT 1, 2, and 3

Since 1986 the SPOT family of satellites has been orbiting the Earth and has already taken more than 10 million high quality images. SPOT 1 was launched with the last Ariane 1 rocket on February 22, 1986. Two days later, the 1800 kg SPOT 1 transmitted its first image with a spatial resolution of 10 or 20 meters. SPOT 2 joined SPOT 1 in orbit on January 22, 1990, on the Ariane 4 maiden flight, and SPOT 3 followed on September 26, 1993, also on an Ariane 4.

The satellite loads were identical, each including two identical HRV (High Resolution Visible) imaging instruments that were able to operate in two modes, either simultaneously or individually. The two spectral modes are panchromatic and multispectral. The panchromatic band has a resolution of 10 meters, and the three multispectral bands (G, R, NIR) have resolutions of 20 metres. They have a scene size of 3600 km² and a revisit interval of one to four days, depending on the latitude.

Because the orbit of SPOT 1 was lowered in 2003, it will gradually lose altitude and break up naturally in the atmosphere. Deorbiting of SPOT 2, in accordance with IADC (Inter-Agency Space Debris Coordination Committee), commenced in mid-July 2009 for a period of two weeks, with a final burn on 29 July 2009. SPOT 3 is no longer functioning, due to problems with its stabilization system.

SPOT 4

SPOT 4 launched March 24, 1998 and stopped functioning July, 2013. In 2013, CNES lowered the altitude of SPOT 4 by 2.5 km to put it on a phased orbit with a five-day repeat cycle. On this orbit, SPOT4 was programmed to acquire a time-lapse series of images over 42 sites with a five days revisit period from February to end of May 2013. The data set it produced is aimed at helping future users of the Sentinel-2 mission to learn working with time-lapse series. The time-lapse series provided by SPOT4 (Take5) have the same repetitiveness as those that will be delivered by the Sentinel-2 satellites, starting in 2015 and 2016.

SPOT 5

SPOT 5 was launched on May 4, 2002 and has the goal to ensure continuity of services for customers and to improve the quality of data and images by anticipating changes in market requirements.

SPOT 5 has two high resolution geometrical (HRG) instruments that were deduced from the HRVIR of SPOT 4. They offer a higher resolution of 2.5 to 5 meters in panchromatic mode and 10 meters in multispectral mode (20 metre on short wave infrared 1.58 − 1.75 µm). SPOT 5 also features an HRS imaging instrument operating in panchromatic mode. HRS points forward and backward of the satellite. Thus, it is able to take stereopair images almost simultaneously to map relief.

SCIAMACHY

SCIAMACHY (SCanning Imaging Absorption SpectroMeter for Atmospheric CHartographY; "fighting shadows") was one of ten instruments aboard of ESA's ENVIronmental

SATellite, ENVISAT. It was a satellite spectrometer designed to measure sunlight, transmitted, reflected and scattered by the earth's atmosphere or surface in the ultra-violet, visible and near infrared wavelength region (240 nm - 2380 nm) at moderate spectral resolution (0.2 nm - 1.5 nm). SCIAMACHY was built by Netherlands and Germany at TNO/TPD, SRON and Dutch Space.

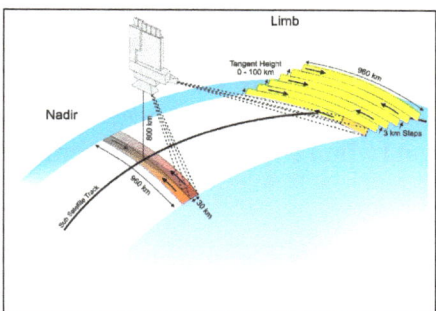

SCIAMACHY, Nadir and Limb scanning.

Launch and Termination

SCIAMACHY, aboard the ENVISAT satellite, was launched by ESA (European Space Agency) from Kourou, French Guiana, in March 2002. ENVISAT's mission was ended in May 2012, after loss of contact one month earlier.

Operation

The absorption, reflection and scattering characteristics of the atmosphere were determined by measuring the extraterrestrial solar irradiance and the upwelling radiance observed in different viewing geometries. The ratio of extraterrestrial irradiance and the upwelling radiance can be inverted to provide information about the amounts and distribution of important atmospheric constituents, which absorb or scatter light, and the spectral reflectance (or albedo) of the Earth's surface.

Purpose

SCIAMACHY was conceived to improve global knowledge and understanding of a variety of issues of importance for the chemistry and physics of the Earth's atmosphere (troposphere, stratosphere and mesosphere) and potential changes resulting from either anthropogenic behavior or natural phenomena.

Weather Satellite

Weather monitoring and forecasting was one of the first civilian (as opposed to military) applications of satellite remote sensing, dating back to the first true weather satellite, TIROS-1 (Television and Infrared Observation Satellite - 1), launched in 1960 by the United States. Several other weather satellites were launched over the next five years, in near-polar orbits, providing repetitive coverage of global weather patterns. In 1966, NASA (the U.S. National Aeronautics and Space Administration) launched the geostationary Applications Technology Satellite (ATS-1) which provided hemispheric images of the Earth's surface and cloud cover every half hour. For the first time, the development and movement of weather systems could be routinely monitored. Today, several countries operate weather, or meteorological satellites to monitor weather conditions around the globe. Generally speaking, these satellites use sensors which have fairly coarse spatial resolution (when compared to systems for observing land) and provide large areal coverage.

Their temporal resolutions are generally quite high, providing frequent observations of the Earth's surface, atmospheric moisture, and cloud cover, which allows for near-continuous monitoring of global weather conditions, and hence - forecasting. Here we review a few of the representative satellites/sensors used for meteorological applications.

GOES

The GOES (Geostationary Operational Environmental Satellite) System is the follow-up to the ATS series. They were designed by NASA for the National Oceanic and Atmospheric Administration (NOAA) to provide the United States National Weather Service with frequent, small-scale imaging of the Earth's surface and cloud cover. The GOES series of satellites have been used extensively by meteorologists for weather monitoring and forecasting for over 20 years. These satellites are part of a global network of meteorological satellites spaced at approximately 70° longitude intervals around the Earth in order to provide near-global coverage. Two GOES satellites, placed in geostationary orbits 36000 km above the equator, each view approximately one-third of the Earth. One is situated at 75°W longitude and monitors North and South America and most of the Atlantic Ocean. The other is situated at 135°W longitude and monitors North America and the Pacific Ocean basin. Together they cover from 20°W to 165°E longitude. This GOES image covers a portion of the southeastern United States, and the adjacent ocean areas where many severe storms originate and develop. This image shows Hurricane Fran approaching the southeastern United States and the Bahamas in September of 1996.

Two generations of GOES satellites have been launched, each measuring emitted and reflected radiation from which atmospheric temperature, winds, moisture, and cloud cover can be derived. The first generation of satellites consisted of GOES-1 through GOES-7. Due to their design, these satellites were capable of viewing the Earth only a small percentage of the time (approximately five per cent). The second generation of satellites began with GOES-8 and has numerous technological improvements over the first series. They provide near-continuous observation of the Earth allowing more frequent imaging (as often as every 15 minutes). This increase in temporal resolution coupled with improvements in the spatial and radiometric resolution of the sensors provides timelier information and improved data quality for forecasting meteorological conditions.

GOES-8 and the other second generation GOES satellites have separate imaging and sounding instruments. The imager has five channels sensing visible and infrared reflected and emitted solar radiation. The infrared capability allows for day and night imaging. Sensor pointing and scan selection capability enable imaging of an entire hemisphere, or small-scale imaging of selected areas. The latter allows meteorologists to monitor specific weather trouble spots to assist in improved short-term forecasting. The imager data are 10-bit radiometric resolution, and can be transmitted directly to local user terminals on the Earth's surface. The accompanying table describes the individual bands, their spatial resolution, and their meteorological applications.

GOES Bands			
Band	Wavelength Range (µm)	Spatial Resolution	Application
1	0.52 - 0.72 (visible)	1 km	Cloud, pollution, and haze detection; severe storm identification.
2	3.78 - 4.03 (shortwave IR)	4 km	Identification of fog at night; discriminating water clouds and snow or ice clouds during daytime; detecting fires and volcanoes; night time determination of sea surface temperatures.
3	6.47 - 7.02 (upper level water vapour)	4 km	Estimating regions of mid-level moisture content and advection; tracking mid-level atmospheric motion.
4	10.2 - 11.2 (longwave IR)	4 km	Identifying cloud-drift winds, severe storms, and heavy rainfall.
5	11.5 - 12.5 (IR window sensitive to water vapour)	4 km	Identification of low-level moisture; determination of sea surface temperature; detection of airborne dust and volcanic ash.

The 19 channel sounder measures emitted radiation in 18 thermal infrared bands and reflected radiation in one visible band. These data have a spatial resolution of 8 km and 13-bit radiometric resolution. Sounder data are used for surface and cloud-top temperatures, multi-level moisture profiling in the atmosphere, and ozone distribution analysis.

NOAA AVHRR

NOAA is also responsible for another series of satellites which are useful for meteorological, as well as other, applications. These satellites, in sun-synchronous, near-polar orbits (830-870 km above the Earth), are part of the Advanced TIROS series and provide complementary information to the geostationary meteorological satellites (such as GOES). Two satellites, each providing global coverage, work together to ensure that data for any region of the Earth is no more than six hours old. One satellite crosses the equator in the early morning from north-to-south while the other crosses in the afternoon.

The primary sensor on board the NOAA satellites, used for both meteorology and small-scale Earth observation and reconnaissance, is the Advanced Very High Resolution Radiometer (AVHRR). The AVHRR sensor detects radiation in the visible, near and mid infrared, and thermal infrared portions of the electromagnetic spectrum, over a swath width of 3000 km. The accompanying table, outlines the AVHRR bands, their wavelengths and spatial resolution (at swath nadir), and general applications of each.

NOAA AVHRR Bands			
Band	Wavelength Range (μm)	Spatial Resolution	Application
1	0.58 - 0.68 (red)	1.1 km	Cloud, snow, and ice monitoring.
2	0.725 - 1.1 (near IR)	1.1 km	Water, vegetation, and agriculture surveys.
3	3.55 -3.93 (mid IR)	1.1 km	Sea surface temperature, volcanoes, and forest fire activity.
4	10.3 - 11.3 (thermal IR)	1.1 km	Sea surface temperature, soil moisture.
5	11.5 - 12.5 (thermal IR)	1.1 km	Sea surface temperature, soil moisture.

AVHRR data can be acquired and formatted in four operational modes, differing in resolution and method of transmission. Data can be transmitted directly to the ground and viewed as data are collected, or recorded on board the satellite for later transmission and processing. The accompanying table describes the various data formats and their characteristics.

AVHRR Data Formats		
Format	Spatial Resolution	Transmission and Processing
APT (Automatic Picture Transmission)	4 km	Low-resolution direct transmission and display.

HRPT (High Resolution Picture Transmission)	1.1 km	Full-resolution direct transmission and display.
GAC (Global Area Coverage)	4 km	Low-resolution coverage from recorded data.
LAC (Local Area Coverage)	1.1 km	Selected full-resolution local area data from recorded data.

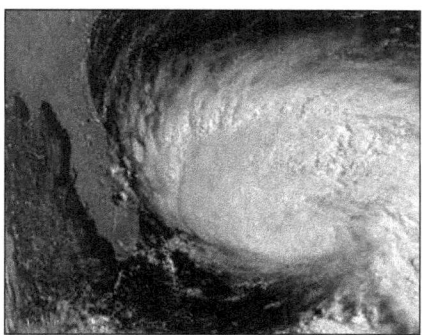

Although AVHRR data are widely used for weather system forecasting and analysis, the sensor is also well-suited to observation and monitoring of land features. AVHRR has much coarser spatial resolution than other typical land observations sensors (discussed in the next section), but is used extensively for monitoring regional, small-scale phenomena, including mapping of sea surface temperature, and natural vegetation and crop conditions. Mosaics covering large areas can be created from several AVHRR data sets allowing small scale analysis and mapping of broad vegetation cover.

Other Weather Satellites

The United States operates the DMSP (Defense Meteorological Satellite Program) series of satellites which are also used for weather monitoring. These are near-polar orbiting satellites whose Operational Linescan System (OLS) sensor provides twice daily coverage with a swath width of 3000 km at a spatial resolution of 2.7 km. It has two fairly broad wavelength bands: a visible and near infrared band (0.4 to 1.1 μm) and a thermal infrared band (10.0 to 13.4 μm). An interesting feature of the sensor is its ability to acquire visible band night time imagery under very low illumination conditions. With this sensor, it is possible to collect striking images of the Earth showing (typically) the night time lights of large urban centres.

There are several other meteorological satellites in orbit, launched and operated by other countries, or groups of countries. These include Japan, with the GMS satellite series, and the consortium of European communities, with the Meteosat satellites. Both are geostationary satellites situated above the equator over Japan and Europe, respectively. Both provide half-hourly imaging of the Earth similar to GOES. GMS has two bands: 0.5 to 0.75 μm (1.25 km resolution), and 10.5 to 12.5 μm (5 km resolution). Meteosat has three bands: visible band (0.4 to 1.1 μm; 2.5 km resolution), mid-IR (5.7 to 7.1 μm; 5 km resolution), and thermal IR (10.5 to 12.5 μm; 5 km resolution).

Weather Satellite

Weather satellites carry instruments called radiometers (not cameras) that scan the Earth to form images. These instruments usually have some sort of small telescope or antenna, a scanning mechanism, and one or more detectors that detect either visible, infrared, or microwave radiation for the purpose of monitoring weather systems around the world.

The measurements these instruments make are in the form of electrical voltages, which are digitized and then transmitted to receiving stations on the ground. The data are then relayed to various weather forecast centers around the world, and are made available over the internet in the form of images. Because weather changes quickly, the time from satellite measurement to image availability can be less than a minute.

Most of the satellites and instruments they carry are designed to operate for 3 to 7 years, although many of them last much longer than that.

Weather satellites are put into one of two kinds of orbits around the Earth, each of which has advantages (and disadvantages) for weather monitoring. The first is a "geostationary" orbit, with the satellite at a very high altitude (about 22,500 miles) and orbiting over the equator at the same rate that the Earth turns. This allows the satellite to view the same geographic area continuously.

For instance, GOES-East and GOES-West provide coverage of much of the Western Hemisphere, from the western coast of Africa to the West Pacific, and the Arctic to the Antarctic. The European Space Agency's Meteosat satellite provides coverage of Europe and Africa.

The disadvantages of a geostationary orbit are (1) its very high altitude, which requires elaborate telescopes and precise scanning mechanisisms in order to image the Earth at high resolution (currently, 1 km at best); and (2) only a portion of the Earth can be viewed.

The other orbit type is called near-polar, sun-synchronous (or just "polar"), where the satellite is put into a relatively low altitude orbit (around 500 miles) that carries the satellite near the North Pole and the South Pole approximately every 100 minutes. Unlike the geostationary orbit, the polar orbit allows complete Earth coverage as the Earth turns beneath it.

These orbits are "sun-synchronous", allowing the satellite to measure the same location on the Earth twice each day at the same local time. Of course, the disadvantage of this orbit is that the satellite can image a particular location only every 12 hours, rather than continuously as in the case of the geostationary satellite. To offset this disadvantage, two satellites put into orbits at different sun-synchronous times have allowed up to 6 hourly monitoring.

But because of the lower altitude (500 miles rather than 22,000 miles), the instruments the polar-orbiting satellite carries to image the Earth do not have to be as elaborate in order to achieve the same ground resolution. Also, the lower orbit allows microwave radiometers to be used, which must have relatively large antennas in order to achieve ground resolutions fine enough to be useful. The advantage of microwave radiometers is their ability to measure through clouds to sense precipitation, temperature in different layers of the atmosphere, and surface characteristics like ocean surface winds.

Because of their global coverage, some of the measurements from polar orbiting satellites are put into computerized weather forecast models, which are the basis for weather forecasting.

References

- "50 years of Earth Observation". 2007: A Space Jubilee. European Space Agency. October 3, 2007. Retrieved 2008-03-20

- Satellite-remote-sensing, earth-and-planetary-sciences, topics: sciencedirect.com, Retrieved 1 July, 2020

- Mcgovern, Eugene A.; Holden, Nicholas M.; Ward, Shane M.; Collins, James F. (January 2000). "Remotely sensed satellite imagery as an information source for industrial peatlands management". Resources, Conservation and Recycling. 28 (1–2): 67–83. Doi:10.1016/s0921-3449(99)00034-8. ISSN 0921-3449

- Sattech, RS, TG: ciesin.org, Retrieved 2 August, 2020

- Friedel, Michael J.; Buscema, Massimo; Vicente, Luiz Eduardo; Iwashita, Fabio; Koga-Vicente, Andréa (2017-07-11). "Mapping fractional landscape soils and vegetation components from Hyperion satellite imagery using an unsupervised machine-learning workflow". International Journal of Digital Earth. 11 (7): 670–690

- Weather-satellitessensors, satellites-sensors, remote-sensing-tutorials, satellite-imagery-air-photos, maps-tools-publications: nrcan.gc.ca, Retrieved 3 January, 2020

- Harland, David M.; Lorenz, Ralph D. (2006) (2005). Space Systems Failures: Disasters and Rescues of Satellites, Rocket and Space Probes. Springer Science+Business Media. P. 107. ISBN 0-387-21519-0

- How-do-weather-satellites-work, weatherquestions: weatherstreet.com, Retrieved 4 February, 2020

- Kramer, Miriam (28 May 2015). "The life and death of Ikonos, a pioneering commercial satellite". Mashable. Retrieved 14 January 2018

3

Geographic Information System and Global Positioning System

Geographic information system deals with capturing, storing, manipulating and managing all types of spatial and geographical data. Global positioning system is a navigation system that consists of a network of 24 satellites for determining the ground position of an object. This chapter has been carefully written to provide an easy understanding of geographic information system and global positioning system.

Geographic Information System

A geographic information system (GIS) is a computer system for capturing, storing, checking, and displaying data related to positions on Earth's surface. By relating seemingly unrelated data, GIS can help individuals and organizations better understand spatial patterns and relationships.

GIS technology is a crucial part of spatial data infrastructure, which the White House defines as "the technology, policies, standards, human resources, and related activities necessary to acquire, process, distribute, use, maintain, and preserve spatial data."

GIS can use any information that includes location. The location can be expressed in many different ways, such as latitude and longitude, address, or ZIP code.

Many different types of information can be compared and contrasted using GIS. The system can include data about people, such as population, income, or education level. It can include information about the landscape, such as the location of streams, different kinds of vegetation, and different kinds of soil. It can include information about the sites of factories, farms, and schools, or storm drains, roads, and electric power lines.

With GIS technology, people can compare the locations of different things in order to discover how they relate to each other. For example, using GIS, a single map could include sites that produce pollution, such as factories, and sites that are sensitive to pollution, such as wetlands and rivers. Such a map would help people determine where water supplies are most at risk.

Data Capture

Data Formats

GIS applications include both hardware and software systems. These applications may include cartographic data, photographic data, digital data, or data in spreadsheets.

Cartographic data are already in map form, and may include such information as the location of rivers, roads, hills, and valleys. Cartographic data may also include survey data and mapping information that can be directly entered into a GIS.

Photographic interpretation is a major part of GIS. Photo interpretation involves analyzing aerial photographs and assessing the features that appear.

Digital data can also be entered into GIS. An example of this kind of information is computer data collected by satellites that show land use—the location of farms, towns, and forests.

Remote sensing provides another tool that can be integrated into a GIS. Remote sensing includes imagery and other data collected from satellites, balloons, and drones.

Finally, GIS can also include data in table or spreadsheet form, such as population demographics. Demographics can range from age, income, and ethnicity to recent purchases and internet browsing preferences.

GIS technology allows all these different types of information, no matter their source or original format, to be overlaid on top of one another on a single map. GIS uses location as the key index variable to relate these seemingly unrelated data.

Putting information into GIS is called data capture. Data that are already in digital form, such as most tables and images taken by satellites, can simply be uploaded into GIS. Maps, however, must first be scanned, or converted to digital format.

The two major types of GIS file formats are raster and vector. Raster formats are grids of cells or pixels. Raster formats are useful for storing GIS data that vary, such as elevation or satellite imagery. Vector formats are polygons that use points (called nodes) and lines. Vector formats are useful for storing GIS data with firm borders, such as school districts or streets.

Spatial Relationships

GIS technology can be used to display spatial relationships and linear networks. Spatial relationships may display topography, such as agricultural fields and streams.

They may also display land-use patterns, such as the location of parks and housing complexes.

Linear networks, sometimes called geometric networks, are often represented by roads, rivers, and public utility grids in a GIS. A line on a map may indicate a road or highway. With GIS layers, however, that road may indicate the boundary of a school district, public park, or other demographic or land-use area. Using diverse data capture, the linear network of a river may be mapped on a GIS to indicate the stream flow of different tributaries.

GIS must make the information from all the various maps and sources align, so they fit together on the same scale. A scale is the relationship between the distance on a map and the actual distance on Earth.

Often, GIS must manipulate data because different maps have different projections. A projection is the method of transferring information from Earth's curved surface to a flat piece of paper or computer screen. Different types of projections accomplish this task in different ways, but all result in some distortion. To transfer a curved, three-dimensional shape onto a flat surface inevitably requires stretching some parts and squeezing others.

A world map can show either the correct sizes of countries or their correct shapes, but it can't do both. GIS takes data from maps that were made using different projections and combines them so all the information can be displayed using one common projection.

GIS Maps

Once all the desired data have been entered into a GIS system, they can be combined to produce a wide variety of individual maps, depending on which data layers are included. One of the most common uses of GIS technology involves comparing natural features with human activity.

For instance, GIS maps can display what man-made features are near certain natural features, such as which homes and businesses are in areas prone to flooding.

GIS technology also allows users to "dig deep" in a specific area with many kinds of information. Maps of a single city or neighborhood can relate such information as average income, book sales, or voting patterns. Any GIS data layer can be added or subtracted to the same map.

GIS maps can be used to show information about numbers and density. For example, GIS can show how many doctors there are in a neighborhood compared with the area's population.

With GIS technology, researchers can also look at change over time. They can use satellite data to study topics such as the advance and retreat of ice cover in polar regions, and how that coverage has changed through time. A police precinct might study changes in crime data to help determine where to assign officers.

One important use of time-based GIS technology involves creating time-lapse photography that shows processes occurring over large areas and long periods of time. For example, data showing the movement of fluid in ocean or air currents help scientists better understand how moisture and heat energy move around the globe.

GIS technology sometimes allows users to access further information about specific areas on a map. A person can point to a spot on a digital map to find other information stored in the GIS about that location. For example, a user might click on a school to find how many students are enrolled, how many students there are per teacher, or what sports facilities the school has.

GIS systems are often used to produce three-dimensional images. This is useful, for example, to geologists studying earthquake faults.

GIS technology makes updating maps much easier than updating maps created manually. Updated data can simply be added to the existing GIS program. A new map can then be printed or displayed on screen. This skips the traditional process of drawing a map, which can be time-consuming and expensive.

GIS Software

Geographic Resources Analysis Support System

Graphic Resources Analysis Support System (commonly termed GRASS GIS) is a geographic information system (GIS) software suite used for geospatial data management and analysis, image processing, producing graphics and maps, spatial and temporal modeling, and visualizing. It can handle raster, topological vector, image processing, and graphic data.

GRASS GIS contains over 350 modules to render maps and images on monitor and paper; manipulate raster and vector data including vector networks; process multispectral image data; and create, manage, and store spatial data.

It is licensed and released as free and open-source software under the GNU General Public License (GPL). It runs on multiple operating systems, including OS X, Windows and Linux. Users can interface with the software features through a graphical user interface (GUI) or by *plugging into* GRASS via other software such as QGIS. They can also interface with the modules directly through a bespoke shell that the application launches or by calling individual modules directly from a standard shell. The latest stable release version (LTS) is GRASS GIS 7, which is available since 2015.

The GRASS Development Team is a multinational group consisting of developers at many locations. GRASS is one of the eight initial Software Projects of the Open Source Geospatial Foundation.

Architecture

GRASS supports raster and vector data in two and three dimensions. The vector data model is topological, meaning that areas are defined by boundaries and centroids; boundaries cannot overlap within one layer. In contrast, OpenGIS Simple Features, defines vectors more freely, much as a non-georeferenced vector illustration program does.

GRASS is designed as an environment in which tools that perform specific GIS computations are executed. Unlike GUI-based application software, the GRASS user is presented with a Unix shell containing a modified environment that supports execution of GRASS commands, termed modules. The environment has a state that includes parameters such as the geographic region covered and the map projection in use. All GRASS modules read this state and additionally are given specific parameters (such as input and output maps, or values to use in a computation) when executed. Most GRASS modules and abilities can be operated via a graphical user interface (provided by a GRASS module), as an alternative to manipulating geographic data in a shell.

The GRASS distribution includes over 350 core modules. Over 100 add-on modules created by users are offered on its website. The libraries and core modules are written in C. Other modules are written in C, C++, Python, Unix shell, Tcl, or other scripting languages. The modules are designed under the Unix philosophy and hence can be combined using Python or shell scripting to build more complex or specialized modules, by users, without knowledge of C programming.

There is cooperation between the GRASS and Quantum GIS (QGIS) projects. Recent versions of QGIS can be executed within the GRASS environment, allowing QGIS to be used as a user-friendly graphical interface to GRASS that more closely resembles other graphical GIS software than does the shell-based GRASS interface.

Another project exists to re-implement GRASS in Java as *JGRASS*.

Integrated Land and Water Information System

Integrated Land and Water Information System (ILWIS) is a geographic information system (GIS) and remote sensing software for both vector and raster processing. Its features include digitizing, editing, analysis and display of data, and production of quality maps. ILWIS was initially developed and distributed by ITC Enschede (International Institute for Geo-Information Science and Earth Observation) in the Netherlands for use by its researchers and students. Since 1 July 2007, it has been released as free software under the terms of the GNU General Public License. Having been used by many students, teachers and researchers for more than two decades, ILWIS is one of the most user-friendly integrated vector and raster software programmes currently available. ILWIS has some very powerful raster analysis modules, a high-precision and flexible vector and point digitizing module, a variety of very practical tools, as well as a great variety of user guides and training modules all available for downloading. The current version is ILWIS 3.8.1. Similar to the GRASS GIS in many respects, ILWIS is currently available natively only on Microsoft Windows. However, a Linux Wine manual has been released.

Features

ILWIS uses GIS techniques that integrate image processing capabilities, a tabular database and conventional GIS characteristics. The major features include:

- Integrated raster and vector design.

- On-screen digitizing.

- Comprehensive set of image processing and remote sensing tools like extensive set of filters, resampling, aggregation, classifications. etc.

- Orthophoto, image georeferencing, transformation and mosaicking.

- Advanced modeling and spatial data analysis.

- 3D visualization with interactive zooming, rotation and panning. "Height" information can be added from multiple types of sources and isn't limited to DEM information.

- Animations of spatial temporal data stacks with the possibility of synchronizartion between different animations.

- Rich map projection and geographic coordinate system library. Optionally custom coordinate systems and on the fly modifications can be added.

- Geostatistical analyses, with Kriging for improved interpolation.

- Import and export using the GDAL/OGR library.

- Advanced data management.

- Stereoscopy tools - To create a stereo pair from two aerial photographs.

- Transparencies at many levels (whole maps, selections, individual elements or properties) to combine different data sources in a comprehensive way.

- Various interfactive diagramming options: Profile, Cross section visualization, Hovmoller diagrams.

- Interactive value dependent presentation of maps (stretching, representation).

- Hydrologic flow operations.

- Surface energy balance operations through the SEBS module.

- GARtrip import - Map Import allows the import of GARtrip Text files with GPS data.

- Spatial Multiple Criteria Evaluation (SMCE).

- Space time Cube: Interactive visualization of multiple attribute spatial temporal data.

- DEM operations including iso line generation.

- Variable Threshold Computation, to help preparing a threshold map for drainage network extraction.

- Horton Statistics, to calculate the number of streams, the average stream length, the average area of catchments for Strahler stream orders.

- Georeference editors.

QGIS

GIS (previously known as Quantum GIS) is a free and open-source cross-platform desktop geographic information system (GIS) application that supports viewing, editing, and analysis of geospatial data.

Functionality

QGIS functions as geographic information system (GIS) software, allowing users to analyze and edit spatial information, in addition to composing and exporting graphical maps. QGIS supports both raster and vector layers; vector data is stored as either point, line, or polygon features. Multiple formats of raster images are supported, and the software can georeference images.

QGIS supports shapefiles, coverages, personal geodatabases, dxf, MapInfo, PostGIS, and other formats. Web services, including Web Map Service and Web Feature Service, are also supported to allow use of data from external sources.

QGIS integrates with other open-source GIS packages, including PostGIS, GRASS GIS, and MapServer. Plugins written in Python or C++ extend QGIS's capabilities. Plugins can geocode using the Google Geocoding API, perform geoprocessing functions similar to those of the standard tools found in ArcGIS, and interface with PostgreSQL/PostGIS, SpatiaLite and MySQL databases.

Development

Gary Sherman began development of Quantum GIS in early 2002, and it became an incubator project of the Open Source Geospatial Foundation in 2007. Version 1.0 was released in January 2009.

Written in C++, QGIS makes extensive use of the Qt library. In addition to Qt, required dependencies of QGIS include GEOS and SQLite. GDAL, GRASS GIS, PostGIS, and PostgreSQL are also recommended, as they provide access to additional data formats.

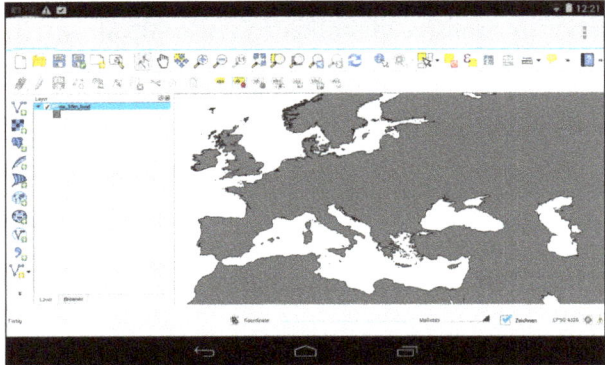

A screenshot from QGIS-Android in 2014.

As of 2017, QGIS is available for multiple operating systems including Mac OS X, Linux, Unix, and Microsoft Windows. A mobile version of QGIS was under development for Android as of 2014.

QGIS can also be used as a graphical user interface to GRASS. QGIS has a small install footprint on the host file system compared to commercial GISs and generally requires less RAM and processing power; hence it can be used on older hardware or running simultaneously with other applications where CPU power may be limited.

QGIS is maintained by volunteer developers who regularly release updates and bug fixes. As of 2012, developers have translated QGIS into 48 languages and the application is used internationally in academic and professional environments. Several companies offer support and feature development services.

Function

Layers

QGIS can display multiple layers containing different sources or depictions of sources.

Preparing Maps

In order to prepare printed map with QGIS, Print Layout is used. It can be used for adding multiple map views, labels, legends, etc.

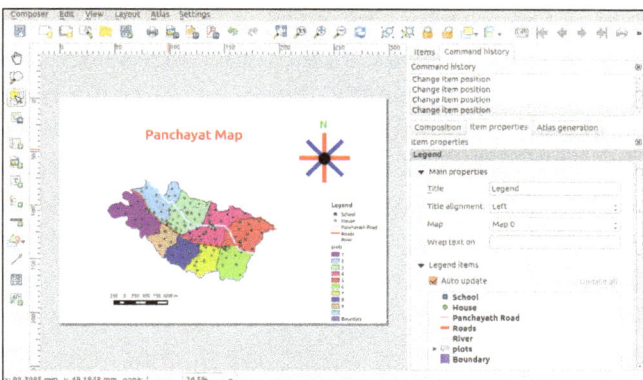

Screenshot of Print Composer.

Adoption

Many public and private organizations have adopted QGIS, including the US National Security Agency, the Austrian state of Vorarlberg, and the Swiss cantons of Glarus and Solothurn.

TerraLook

TerraLook is a free satellite image viewing tool. It provides access to satellite images for users that lack prior experience with remote sensing or Geographic Information System (GIS) technology. It does this by combining user-created collections of images on themes with a set of simple visualization and analysis tools, allowing the user to explore the data and employ it for useful purposes. The images include recent high-resolution ASTER images, and Landsat images from three historical periods going back to the early 70s. This historical data can support change analysis.

Access to satellite images has been largely limited to science communities with necessary resources like expensive software, tools and expertise. However, other types of disciplines including conservation, development planning, education, urban studies, or disaster planning and response could benefit from the ability to view and analyze chronologically and georeferenced satellite images. This type of data could be of particular use in developing countries that may have less capacity to purchase or work with remote sensing technology. This situation has resulted in the under-utilization of valuable data. Fortunately, these access hurdles can be overcome with tools like TerraLook. Google Earth, for example, has had a tremendous impact on the availability of image and vector data. However, it can be inappropriate for addressing certain types of questions such as those pertaining to change studies important to conservationists. Although Google Earth is very robust in its resolution and layering capability, bandwidth, time-series capability, and image processing options can be limited.

TerraLook consists of two parts: TerraLook Image Collections based on themes, and the TerraLook Software to work with them. Image Collections. Users can create their own custom collections of images by visiting the USGS TerraLook website. Using the USGS Global Visualization Viewer (GloVis), the images are selected, the collection built, and the user notified that it is available for download. Also, many standard collections have been archived and are available for download from the ASTER TerraLook website. Typically, these consist of country collections, containing full-country coverage with the historical Landsat images from 1975, 1990, and 2000, plus ASTER coverage of the Protected Areas of the country.

TerraLook Software

TerraLook Software is an Open Source Viewer/Toolkit. The TerraLook Software was formerly known as the Protected Area Archive.

This tool includes the following capabilities, images from four sensors (satellites):

ASTER: 2000 to present, Landsat: c. 2000, Landsat: c. 1990, Landsat: c. 1975.

- Image Finder: Displays footprint (satellite) of images and outlines of protected areas so users can easily find and display images of interest.

- Roam and Zoom: To find and see features of interest within an image.

Measure Distance and Area

- Observing Change: Image "flicker" and side-by-side viewing.

- Overlay Manager: To create, edit, and control overlays, such as country and PA boundaries, add user-supplied overlays, read in shapefiles from a file, enter points into a table.

- Image Annotation: Adding text, lines, polygons, etc., so an image can be used to tell a story and effectively communicate an issue.

- Image Mosaicking: Combines multiple images into a single larger image.

- Image Enhancement: Adjusts images to bring out desired features.

- Multilingual: English and Spanish versions.

- Advanced Features: Band math, classify image, classify layer, attribute editing, 3-D viewing, more.

Global Positioning System

Global Positioning System is a space-based radio-navigation system that broadcasts highly accurate navigation pulses to users on or near Earth. In the United States' Navstar GPS, 24 main satellites in 6 orbits circle Earth every 12 hours. In addition, Russia maintains a constellation called GLONASS (Global Navigation Satellite System), and in 2007 the European Union approved financing for the launch of 30 satellites to form its own version of GPS, known as Galileo, which is projected to be fully operational by 2020. China launched two satellites in 2000 and another in 2003 as part of a local navigation system first known as BeiDou ("Big Dipper"). In 2006 China, which had a limited participation in Galileo, announced plans to expand BeiDou to a full GPS service known as the BeiDou Navigation System. In 2007 China began launching a series of second-generation satellites, known as BeiDou-2, or Compass.

A GPS receiver operated by a user on Earth measures the time it takes radio signals to travel from four or more satellites to its location, calculates the distance to each satellite, and from this calculation determines the user's longitude, latitude, and altitude. The basic civilian service will locate a receiver within 10 metres (33 feet) of its true location, though various augmentation techniques can be used to pinpoint the location within less than 1 cm (0.4 inch). With such accuracy and the ubiquity of the service, GPS has evolved far beyond its original military purpose and has created a revolution in personal and commercial navigation. Battlefield missiles and artillery projectiles use GPS signals to determine their positions and velocities, but so do the U.S. space shuttle and the International Space Station as well as commercial jetliners and private airplanes. Ambulance fleets, family automobiles, and railroad locomotives benefit from GPS positioning, which also serves farm tractors, ocean liners, hikers, and even golfers. Many GPS receivers are no larger than a pocket calculator and are powered by disposable batteries, while GPS computer chips the size of a baby's fingernail have been installed in wristwatches, cellular telephones, and personal digital assistants.

Triangulation

The principle behind the unprecedented navigational capabilities of GPS is triangulation. To triangulate, a GPS receiver precisely measures the time it takes for a satellite signal to make its brief journey to Earth—less than a tenth of a second. Then it multiplies that time by the speed of a radio wave—300,000 km (186,000 miles) per second—to obtain the corresponding distance between it and the satellite. This puts the receiver somewhere on the surface of an imaginary sphere with a radius equal to its distance from the satellite. When signals from three other satellites are similarly processed, the receiver's built-in computer calculates the point at which all four spheres intersect, effectively determining the user's current longitude, latitude, and altitude. (In theory, three satellites would normally provide an unambiguous three-dimensional fix, but in practice at least four are used to offset inaccuracy in the receiver's clock.) In addition, the receiver calculates current velocity (speed and direction) by measuring the instantaneous Doppler effect shifts created by the combined motion of the same four satellites.

In the Navstar system, each satellite broadcasts its navigation signals on two frequencies—1575.42 megahertz (military/civilian) and 1227.6 megahertz (military). These carrier waves are modulated by two pseudo-random binary pulse trains: a 1-megabit-per-second civilian C/A-code (coarse acquisition code) and a 10-megabit-per-second military P-code (precision code). Three new civilian signals are planned at 1176.45, 1227.6, and 1575.42 MHz. Until 2000, a feature known as selective availability (S/A) intentionally degraded the civilian signal's accuracy; S/A was terminated in part because of safety concerns related to the increasing use of GPS by civilian marine vessels and aircraft. Unaugmented civilian GPS now gives an error variance, for horizontal distances, of 30 metres (100 feet) with a probability of 95 percent—that is, 95 percent of the time the reported location is within 30 metres of the true location. Typical horizontal accuracy is about 10 metres (30 feet; compared with 100 metres [330 feet] with S/A), while vertical accuracy, or altitude, is approximately half as precise. The Doppler effect allows receivers to determine a user's velocity to an accuracy of about 1 metre (3 feet) per second. The unaugmented military signal, meanwhile, has a horizontal error variance of less than 3 metres (10 feet).

Augmentation

Although the travel time of a satellite signal to Earth is only a fraction of a second, much can happen to it in that interval. For example, electrically charged particles in the ionosphere and density variations in the troposphere may act to slow and distort satellite signals. These influences can translate into positional errors for GPS users—a problem that can be compounded by timing errors in GPS receiver clocks. Further errors may be introduced by relativistic time dilations, a phenomenon in which a satellite's clock and a receiver's clock, located in different gravitational fields and traveling at different velocities, tick at different rates. Finally, the single greatest source of error to users of the Navstar system is the lower accuracy of the civilian C/A-code pulse. However, various augmentation methods exist for improving the accuracy of both the military and the civilian systems.

When positional information is required with pinpoint precision, users can take advantage of differential GPS techniques. Differential navigation employs a stationary "base station" that sits at a known position on the ground and continuously monitors the signals being broadcast by GPS satellites in its view. It then computes and broadcasts real-time navigation corrections to nearby roving receivers. Each roving receiver, in effect, subtracts its position solution from the base station's solution, thus eliminating any statistical errors common to the two. The U.S. Coast Guard maintains a network of such base stations and transmits corrections over radio beacons covering most of the United States. Other differential corrections are encoded within the normal broadcasts of commercial radio stations. Farmers receiving these broadcasts have been able to direct their field equipment with great accuracy, making precision farming a common term in agriculture.

Another GPS augmentation technique uses the carrier waves that convey the satellites' navigation pulses to Earth. Because the length of the carrier wave is more than 1,000 times shorter than the basic navigation pulses, this "carrier-aided" approach, under the right circumstances, can reduce navigation errors to less than 1 cm (0.4 inch). The dramatically improved accuracy stems primarily from the shorter length and much greater numbers of carrier waves impinging on the receiver's antenna each second.

Yet another augmentation technique is known as geosynchronous overlays. Geosynchronous overlays employ GPS payloads "piggybacked" aboard commercial communication satellites that are placed in geostationary orbit some 35,000 km (22,000 miles) above Earth. These relatively small payloads broadcast civilian C/A-code pulse trains to ground-based users. The U.S. government is enlarging the Navstar constellation with geosynchronous overlays to achieve improved coverage, accuracy, and survivability. Both the European Union and Japan are installing their own geosynchronous overlays.

The Navstar System

The Navstar GPS system consists of three major segments: the space segment, the control segment, and the user segment. The space component is made up of the Navstar constellation in orbit around Earth. The first satellite was an experimental Block I model launched in 1978. Nine more of these developmental satellites followed over the next decade, and 23 heavier and more-capable Block II production models were sent into space from 1989 to 1993. The launch of the 24th Block II satellite in 1994 completed the GPS constellation, which now consists of two dozen Block II satellites (plus three spares orbiting in reserve) marching in single file in six circular orbits around Earth. The orbits are arranged so that at least five satellites are in view from most points on Earth at all times. Since 1994, newer versions of Block II satellites have been launched to replace older models. The first satellite of Block III is scheduled for launch in 2014.

A typical Block II satellite weighs approximately 900 kg (2,000 pounds) and, with its solar panels extended, is about 17 metres (56 feet) across. Its key elements are the winglike solar arrays that generate electrical power from sunlight, the 12 helical antennas that transmit navigation pulses to users on the ground, and its long, spearlike radio antenna that picks up instructions from control engineers. As a satellite coasts through its 12-hour orbit, its main body pivots continuously and the solar arrays swivel, keeping its navigation antennas pointing toward Earth's centre and its solar arrays aligned perpendicular to the Sun's rays.

The control segment consists of one Master Control Station at a U.S. Air Force base in Colorado and four additional unmanned monitoring stations positioned around the world—Hawaii and Kwajalein Atoll in the Pacific Ocean, Diego Garcia in the Indian Ocean, and Ascension Island in the Atlantic Ocean. Each monitoring station tracks all of the GPS satellites in its view to check for orbital changes. Variations in satellite orbits are caused by gravitational pulls from the Moon and Sun, the nonspherical shape of

Earth, and the pressure of solar radiation. This information is processed at the Master Control Station, and corrected orbital information is quickly relayed back to the satellites via large ground antennas. Every 18 months on average, the satellites within a given ring drift too far from their original configuration and must be nudged back with onboard thrusters fired by ground control.

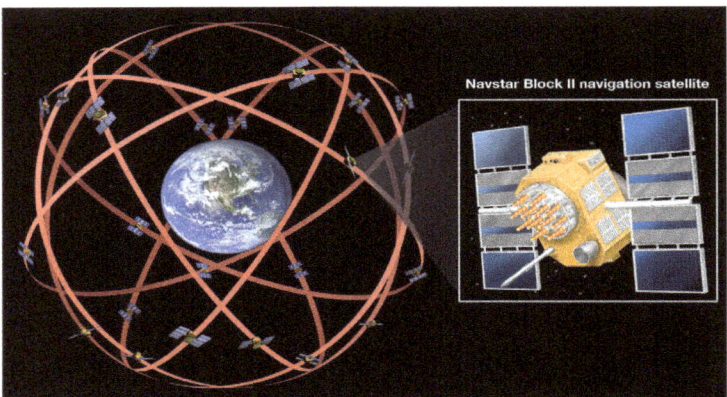

The Navstar navigation system.

The user segment consists of the millions of GPS receivers that pick up and decode the satellite signals. Hundreds of different types of GPS receivers are in use; some are designed for installation in automobiles, trucks, submarines, ships, aircraft, and orbiting satellites, whereas smaller models have been developed for personal navigation.

References

- Neteler, M.; Mitasova, H. (2008). Open Source GIS: a GRASS GIS approach (3rd ed.). New York: Springer. ISBN 978-0-387-35767-6

- Geographic-information-system-gis, encyclopedia: nationalgeographic.org, Retrieved 29 May, 2020

- "ILWIS 3.4 Open". 52°North. 2007-03-27. Archived from the original on 2007-07-07. Retrieved 2007-07-01

- GPS, technology: britannica.com, Retrieved 30 June, 2020

4

Radar and its Types

Radar is referred to the detection system that determines the range, angle and location of objects through the use of radio waves. A few of its types are weather radar, Doppler radar, synthetic-aperture radar, imaging radar, etc. This chapter closely examines the different types of radar to provide an extensive understanding of the subject.

Radar is the electromagnetic sensor used for detecting, locating, tracking, and recognizing objects of various kinds at considerable distances. It operates by transmitting electromagnetic energy toward objects, commonly referred to as targets, and observing the echoes returned from them. The targets may be aircraft, ships, spacecraft, automotive vehicles, and astronomical bodies, or even birds, insects, and rain. Besides determining the presence, location, and velocity of such objects, radar can sometimes obtain their size and shape as well. What distinguishes radar from optical and infrared sensing devices is its ability to detect faraway objects under adverse weather conditions and to determine their range, or distance, with precision.

Radar is an "active" sensing device in that it has its own source of illumination (a transmitter) for locating targets. It typically operates in the microwave region of the electromagnetic spectrum—measured in hertz (cycles per second), at frequencies extending from about 400 megahertz (MHz) to 40 gigahertz (GHz). It has, however, been used at lower frequencies for long-range applications (frequencies as low as several megahertz, which is the HF [high-frequency], or shortwave, band) and at optical and infrared frequencies (those of laser radar,or lidar). The circuit components and other hardware of radar systems vary with the frequency used, and systems range in size from those small enough to fit in the palm of the hand to those so enormous that they would fill several football fields.

Radar underwent rapid development during the 1930s and '40s to meet the needs of the military. It is still widely employed by the armed forces, where many technological advances have originated. At the same time, radar has found an increasing number of important civilian applications, notably air traffic control, weather observation, remote sensing of the environment, aircraft and ship navigation, speed measurement for industrial applications and for law enforcement, space surveillance, and planetary observation.

Fundamentals of Radar

Radar typically involves the radiating of a narrow beam of electromagnetic energy into space from an antenna. The narrow antenna beam scans a region where targets are expected. When a target is illuminated by the beam, it intercepts some of the radiated energy and reflects a portion back toward the radar system. Since most radar systems do not transmit and receive at the same time, a single antenna is often used on a time-shared basis for both transmitting and receiving.

A receiver attached to the output element of the antenna extracts the desired reflected signals and (ideally) rejects those that are of no interest. For example, a signal of interest might be the echo from an aircraft. Signals that are not of interest might be echoes from the ground or rain, which can mask and interfere with the detection of the desired echo from the aircraft. The radar measures the location of the target in range and angular direction. Range, or distance, is determined by measuring the total time it takes for the radar signal to make the round trip to the target and back. The angular direction of a target is found from the direction in which the antenna points at the time the echo signal is received. Through measurement of the location of a target at successive instants of time, the target's recent track can be determined. Once this information has been established, the target's future path can be predicted. In many surveillance radar applications, the target is not considered to be "detected" until its track has been established.

Pulse Radar

The most common type of radar signal consists of a repetitive train of short-duration pulses. The figure shows a simple representation of a sine-wave pulse that might be generated by the transmitter of a medium-range radar designed for aircraft detection. The sine wave in the figure represents the variation with time of the output voltage of the transmitter. The numbers given in parentheses in the figure are meant only to be illustrative and are not necessarily those of any particular radar. They are, however, similar to what might be expected for a ground-based radar system with a range of about 50 to 60 nautical miles (90 to 110 km), such as the kind used for air traffic control at airports. The pulse width is given in the figure as 1 microsecond (10^{-6} second). It should be noted that the pulse is shown as containing only a few cycles of the sine wave; however, in a radar system having the values indicated, there would be 1,000 cycles within the pulse. In the figure the time between successive pulses is given as 1 millisecond (10^{-3} second), which corresponds to a pulse repetition frequency of 1 kilohertz (kHz). The power of the pulse, called the peak power, is taken here to be 1 megawatt. Since a pulse radar does not radiate continually, the average power is much less than the peak power. In this example, the average power is 1 kilowatt. The average power, rather than the peak power, is the measure of the capability of a radar system. Radars have average powers from a few mlli-watts to as much as one or more megawatts, depending on the application.

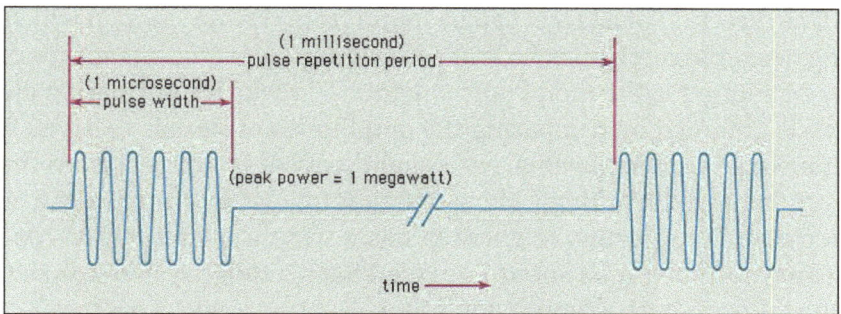

A typical pulse waveform transmitted by radar.

A weak echo signal from a target might be as low as 1 picowatt (10^{-12} watt). In short, the power levels in a radar system can be very large (at the transmitter) and very small (at the receiver).

Another example of the extremes encountered in a radar system is the timing. An air-surveillance radar (one that is used to search for aircraft) might scan its antenna 360 degrees in azimuth in a few seconds, but the pulse width might be about one microsecond in duration. Some radar pulse widths are even of nanosecond (10^{-9} second) duration.

Radar waves travel through the atmosphere at roughly 300,000 km per second (the speed of light). The range to a target is determined by measuring the time that a radar signal takes to travel out to the target and back. The range to the target is equal to $cT/_2$, where c = velocity of propagation of radar energy, and T = round-trip time as measured by the radar. From this expression, the round-trip travel of the radar signal through air is at a rate of 150,000 km per second. For example, if the time that it takes the signal to travel out to the target and back was measured by the radar to be 0.0006 second (600 microseconds), then the range of the target would be 90 km. The ability to measure the range to a target accurately at long distances and under adverse weather conditions is radar's most distinctive attribute. There are no other devices that can compete with radar in the measurement of range.

The range accuracy of a simple pulse radar depends on the width of the pulse: the shorter the pulse, the better the accuracy. Short pulses, however, require wide bandwidths in the receiver and transmitter (since bandwidth is equal to the reciprocal of the pulse width). A radar with a pulse width of one microsecond can measure the range to an accuracy of a few tens of metres or better. Some special radars can measure to an accuracy of a few centimetres. The ultimate range accuracy of the best radars is limited by the known accuracy of the velocity at which electromagnetic waves travel.

Directive Antennas and Target Direction

Almost all radars use a directive antenna—i.e., one that directs its energy in a narrow beam. (The beam width of an antenna of fixed size is inversely proportional to the radar frequency.) The direction of a target can be found from the direction in which the antenna is pointing when the received echo is at a maximum. A precise means for determining the direction of a target is the mono pulse method—in which information about the angle of a target is obtained by comparing the amplitudes of signals received from two or more simultaneous receiving beams, each slightly offset (squinted) from the antenna's central axis. A dedicated tracking radar—one that follows automatically a single target so as to determine its trajectory—generally has a narrow, symmetrical "pencil" beam. (A typical beam width might be about 1 degree.) Such a radar system can determine the location of the target in both azimuth angle and elevation angle. An aircraft-surveillance radar generally employs an antenna that radiates a "fan" beam, one that is narrow in azimuth (about 1 or 2 degrees) and broad in elevation (elevation beam widths of from 20 to 40 degrees or more). A fan beam allows only the measurement of the azimuth angle.

Doppler Frequency and Target Velocity

Radar can extract the Doppler frequency shift of the echo produced by a moving target by noting how much the frequency of the received signal differs from the frequency of the signal that was transmitted. (The Doppler effect in radar is similar to the change in audible pitch experienced when a train whistle or the siren of an emergency vehicle moves past the listener.) A moving target will cause the frequency of the echo signal to increase if it is approaching the radar or to decrease if it is receding from the radar. For example, if a radar system operates at a frequency of 3,000 MHz and an aircraft is moving toward it at a speed of 400 knots (740 km per hour), the frequency of the received echo signal will be greater than that of the transmitted signal by about 4.1 kHz. The Doppler frequency shift in hertz is equal to $3.4 f_o v_r$, where f_o is the radar frequency in gigahertz and v_r is the radial velocity (the rate of change of range) in knots.

Since the Doppler frequency shift is proportional to radial velocity, a radar system that measures such a shift in frequency can provide the radial velocity of a target. The Doppler frequency shift can also be used to separate moving targets from stationary targets even when the echo signal from undesired clutter is much more powerful than the echo from the desired moving targets. A form of pulse radar that uses the Doppler frequency shift to eliminate stationary clutter is called either a moving-target indication (MTI) radar or a pulse Doppler radar, depending on the particular parameters of the signal waveform.

The above measurements of range, angle, and radial velocity assume that the target is a "point-scatter." Actual targets, however, are of finite size and can have distinctive shapes. The range profile of a finite-sized target can be determined if the range resolution of the radar is small compared with the target's size in the range dimension. (The range resolution of a radar, given in units of distance, is a measure of the ability of a radar to separate two closely spaced echoes.) Some radars can have resolutions much smaller than one metre, which is quite suitable for determining the radial size and profile of many targets of interest.

The resolution in angle, or cross range, that can be obtained with conventional antennas is poor compared with that which can be obtained in range. It is possible, however, to achieve good resolution in angle by resolving in Doppler frequency (i.e., separating one Doppler frequency from another). If the radar is moving relative to the target (as when the radar is on an aircraft and the target is the ground), the Doppler frequency shift will be different for different parts of the target. Thus, the Doppler frequency shift can allow the various parts of the target to be resolved. The resolution in cross range derived from the Doppler frequency shift is far better than that achieved with a narrow-beam antenna. It is not unusual for the cross-range resolution obtained from Doppler frequency to be comparable to that obtained in the range dimension.

Radar Imaging

Radar can distinguish one kind of target from another (such as a bird from an aircraft), and some systems are able to recognize specific classes of targets (for example, a commercial airliner as opposed to a military jet fighter). Target recognition is accomplished by measuring the size and speed of the target and by observing the target with high resolution in one or more dimensions. Propellers and jet engines modify the radar echo from aircraft and can assist in target recognition. The flapping of the wings of a bird in flight produces a characteristic modulation that can be used to recognize that a bird is present or even to distinguish one type of bird from another.

Cross-range resolution obtained from Doppler frequency, along with range resolution, is the basis for synthetic aperture radar (SAR). SAR produces an image of a scene that is similar, but not identical, to an optical photograph. One should not expect the image seen by radar "eyes" to be the same as that observed by optical eyes. Each provides different information. Radar and optical images differ because of the large difference in the frequencies involved; optical frequencies are approximately 100,000 times higher than radar frequencies.

SAR can operate from long range and through clouds or other atmospheric effects that limit optical and infrared imaging sensors. The resolution of a SAR image can be made independent of range, an advantage over passive optical imaging where the resolution worsens with increasing range. Synthetic aperture radars that map areas of the Earth's surface with resolutions of a few metres can provide information about the nature of the terrain and what is on the surface.

A SAR operates on a moving vehicle, such as an aircraft or spacecraft, to image stationary objects or planetary surfaces. Since relative motion is the basis for the Doppler resolution, high resolution (in cross range) also can be accomplished if the radar is stationary and the target is moving. This is called inverse synthetic aperture radar (ISAR). Both the target and the radar can be in motion with ISAR.

Basic Radar System

The figure shows the basic parts of a typical radar system. The transmitter generates the high-power signal that is radiated by the antenna. In a sense, an antenna acts as a "transducer" to couple electromagnetic energy from the transmission line to radiation in space, and vice versa. The duplexer permits alternate transmission and reception with the same antenna; in effect, it is a fast-acting switch that protects the sensitive receiver from the high power of the transmitter.

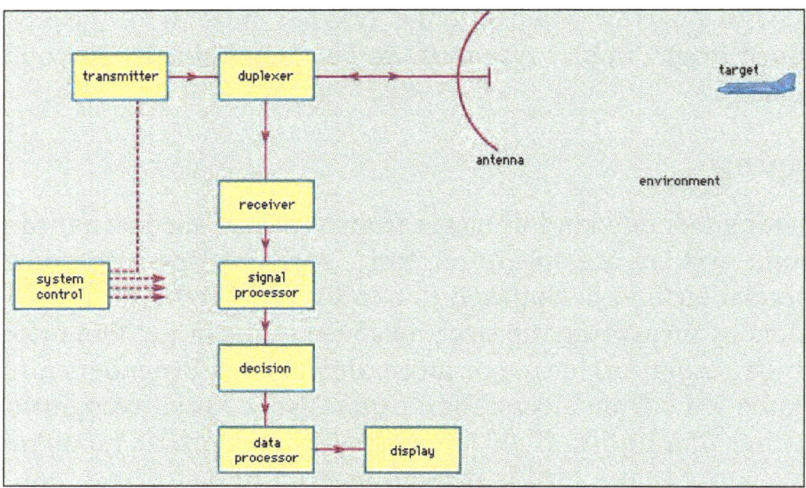

Basic parts of a radar system.

The receiver selects and amplifies radar echoes so that they can be displayed on a television-like screen for the human operator or be processed by a computer. The signal processor separates the signals reflected by possible targets from unwanted clutter. Then, on the basis of the echo's exceeding a predetermined value, a human operator or a digital computer circuit decides whether a target is present.

Once it has been decided that a target is present and its location (in range and angle) has been determined, the track of the target can be obtained by measuring the target location at different times. During the early days of radar, target tracking was performed by an operator marking the location of the target "blip" on the face of a cathode-ray tube (CRT) display with a grease pencil. Manual tracking has been largely replaced by automatic electronic tracking, which can process hundreds or even thousands of target tracks simultaneously.

The system control optimizes various parameters on the basis of environmental conditions and provides the timing and reference signals needed to permit the various parts of the radar to operate effectively as an integrated system.

Antennas

A widely used form of radar antenna is the parabolic reflector, the principle of which is shown in cross section in part A of the figure. A horn antenna (not shown) or other small antenna is placed at the focus of the parabola to illuminate the parabolic surface of the reflector. After being reflected by this surface, the electromagnetic energy is radiated as a narrow beam. A paraboloid, which is generated by rotating a parabola about its axis, forms a symmetrical beam called a pencil beam. A fan beam, one with a narrow beam width in azimuth and a broad beam width in elevation, can be obtained by illuminating an asymmetrical section of the paraboloid. An example of an antenna that produces a fan beam is shown in the photograph.

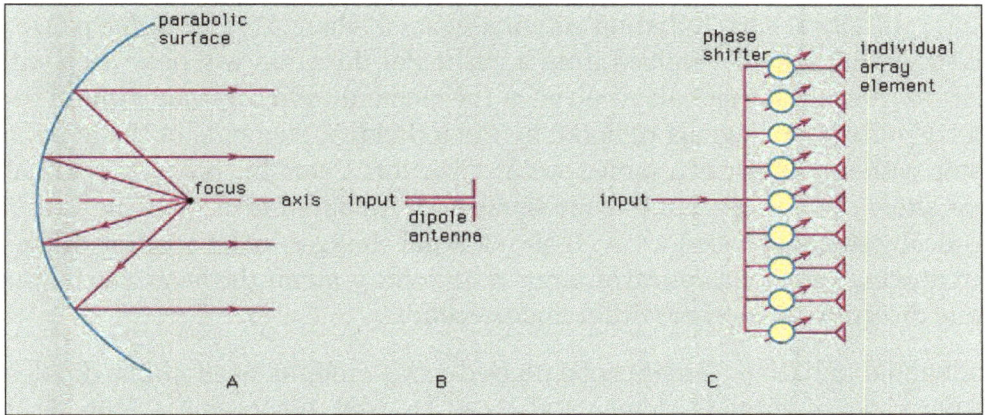

Radar antennas: (A) A parabolic reflector antenna in which the energy radiated from the focus is reflected from the parabolic surface as a narrow beam. (B) A dipole antenna. (C) A phased-array antenna composed of many individual radiating elements.

Reflector antenna for the ASR-9 airport surveillance radar, with an air-traffic-control radar-beacon system (ATCRBS), or Mode S, antenna mounted on top.

The half-wave dipole, whose dimension is one-half of the radar wavelength, is the classic type of electromagnetic antenna. A single dipole is not of much use for radar, since it produces a beamwidth too wide for most applications. Radar requires a narrow beam (a beamwidth of only a few degrees) in order to concentrate its energy on the target and to determine the target location with accuracy. Such narrow beams can be formed by combining many individual dipole antennas so that the signals radiated or received by each elemental dipole are in unison, or in step. (The radar engineer would say that the signals are "in phase" with one another or that they are coherently added together.) This is called a phased-array antenna, the basic principle of which is shown in part C of the figure.

The phase shifters at each radiating antenna-element change (or shift) the phase of the signal, so that all signals received from a particular direction will be in step with one another. As a result, the signals received at the elements add together without theoretical loss. Similarly, all signals radiated by the individual elements of the antenna will be in step with one another in some specific direction. Changing the phase shift at each element alters the direction of the antenna beam. An antenna of this kind is called an electronically steered phased-array. It allows rapid changes in the position of the beam without moving large mechanical structures. In some systems the beam can be changed from one direction to another within microseconds.

The individual radiating elements of a phased-array antenna need not be dipoles; various other types of antenna elements also can be used. For example, slots cut in the side of a waveguide are common, especially at the higher microwave frequencies. In a radar that requires a one-degree pencil-beam antenna, there might be about 5,000 individual radiating elements (the actual number depends on the particular design). The phased-array radar is more complex than radar systems that employ reflector antennas, but it provides capabilities not otherwise available.

Since there are many control points (each individual antenna element) in a phased-array, the radiated beam can be shaped to give a desired pattern to the beam. Controlling the shape of the radiated beam is important when the beam has to illuminate the airspace where aircraft are found but not illuminate the ground, where clutter echoes are produced. Another example is when the stray radiation (called antenna sidelobes) outside the main beam of the antenna pattern must be minimized.

The electronically steered phased-array is attractive for applications that require large antennas or when the beam must be rapidly changed from one direction to another. Satellite surveillance radars and long-range ballistic-missile-detection radars are examples that usually require phased-arrays. The U.S. Army's Patriot battlefield air-defense system and the U.S. Navy's Aegis system for ship air defense also depend on the electronically steered phased-array antenna.

The phased-array antenna is also used in some applications without the phase shifters shown in part C of the figure. The beam is steered by the mechanical movement of the entire antenna. Antennas of this sort are preferred over the parabolic reflector for airborne applications, in land-based air-surveillance radars requiring multiple beams (as in the so-called 3D radars, which measure elevation angle in addition to azimuth and range), and in applications that require ultralow antenna sidelobe radiation.

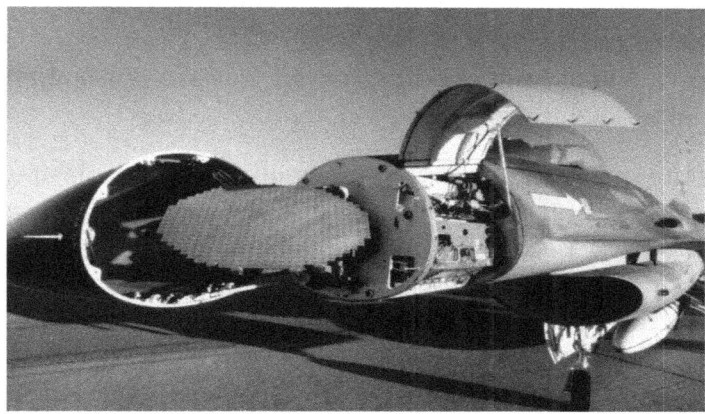

The AN/APG-66 radar in an F-16 fighter aircraft: The mechanically scanned planar phased-array antenna with radiating horizontal slots is 29 inches (74 cm) wide by 19 inches (48 cm) high.

Transmitters

The transmitter of a radar system must be efficient, reliable, not too large in size and weight, and easily maintained, as well as have the wide bandwidth and high power that are characteristic of radar applications. In general, the transmitter must generate low-noise, stable transmissions so that extraneous (unwanted) signals from the transmitter do not interfere with the detection of the small Doppler frequency shift produced by weak moving targets.

The magnetron transmitter has certain limitations, but it continues to be used, for example, in low-average-power applications such as ship navigation radar and airborne weather-avoidance radar. The magnetron is a power oscillator in that it self-oscillates (i.e., generates microwave energy) when voltage is applied. Other radar transmitters usually are power amplifiers in that they take low-power signals at the input and amplify them to high power at the output. This provides stable high-power signals, as the signals to be radiated can be generated with precision at low power.

The klystron amplifier is capable of some of the highest power levels used in radar (many hundreds of kilowatts of average power). It has good efficiency and good stability. The disadvantages of the klystron are that it is usually large and it requires high voltages (e.g., about 90 kilovolts for one megawatt of peak power). At low power the instantaneous bandwidth of the klystron is small, but the klystron is capable of large bandwidth at high peak powers of a few megawatts.

The traveling-wave tube (TWT) is related to the klystron. It has very wide bandwidths at low peak power, but, as the peak power levels are increased to those needed for pulse radar, its bandwidth decreases. As peak power increases, the bandwidths of the TWT and the klystron approach one another.

Solid-state transmitters, such as the transistor, are attractive because of their potential for long life, ease of maintenance, and relatively wide bandwidth. An individual solid-state device generates relatively low power and can be used only when the radar application can be accomplished with low power (as in short-range applications or in the radar altimeter). High power can be achieved, however, by combining the outputs of many individual solid-state devices.

While the solid-state transmitter is easy to maintain and is capable of wide-band operation, it has certain disadvantages. It is much better suited for long pulses (milliseconds) than for short pulses (microseconds). Long pulses can complicate radar operation because signal processing (such as pulse compression) is needed to achieve the desired range resolution. Furthermore, a long-pulse radar generally requires several different pulse widths: a long pulse for long range and one or more shorter, high-energy pulses with less energy to observe targets at the ranges masked when the long pulse is transmitting. (A one-millisecond pulse, for example, masks echoes from 0 to about 80 nautical miles, or 150 km.).

Every kind of transmitter has its disadvantages as well as advantages. In any particular application, the radar engineer must continually search for compromises that give the results desired without too many negative effects that cannot be adequately accommodated.

Receivers

Like most other receivers, the radar receiver is a classic superheterodyne. It has to filter the desired echo signals from clutter and receiver noise that interfere with detection. It also must amplify the weak received signals to a level where the receiver output is large enough to actuate a display or a computer. The technology of the radar receiver is well established and seldom sets a limit on radar performance.

The receiver must have a large dynamic range in situations where it is necessary to detect weak signals in the presence of very large clutter echoes by recognizing the Doppler frequency shift of the desired moving targets. Dynamic range can be loosely described as the ratio of the strongest to the weakest signals that can be handled without significant distortion by a receiver. A radar receiver might be required to detect signals that vary in power by a million to one—and sometimes much more.

In most cases the sensitivity of a radar receiver is determined by the noise generated internally at its input. Because it does not generate much noise of its own, a transistor is usually used as the first stage of a receiver.

Signal and Data Processors

The signal processor is the part of the receiver that extracts the desired target signal from unwanted clutter. It is not unusual for these undesired reflections to be much larger than desired target echoes, in some cases more than one million times larger. Large clutter echoes from stationary objects can be separated from small moving target echoes by noting the Doppler frequency shift produced by the moving targets. Most signal processing is performed digitally with computer technology. Digital processing has significant capabilities in signal processing not previously available with analog methods.

Pulse compression is sometimes included under signal processing. It too benefits from digital technology, but analog processors (e.g., surface acoustic wave delay lines) are used rather than digital methods when pulse compression must achieve resolutions of a few feet or less.

Displays

Although it has its limitations, the cathode-ray tube (CRT) has been the preferred technology for displaying information ever since the early days of radar. There have been, however, considerable improvements in flat-panel displays because of the demands of computers and television. Flat-panel displays occupy less volume and require less power than CRTs, but they also have their limitations. Radar has taken advantage of flat-panel displays and has become increasingly important as a display.

In the early days of radar, an operator decided whether a target was present on the basis of what raw data were displayed. Modern radars, however, present processed information to the operator. Detections are made automatically in the receiver without operator involvement and are then presented on the display to the operator for further action.

A commonly used radar display is the plan position indicator (PPI), which provides a maplike presentation in polar coordinates of range and angle. The display is "dark" except when echo signals are present.

All practical radar displays have been two-dimensional, yet many radars provide more information than can be displayed on the two coordinates of a flat screen. Colour coding of the signal indicated on the PPI is sometimes used to provide additional information about the echo signal. Colour has been employed, for example, to indicate the strength of the echo. Doppler weather radars make good use of colour coding to indicate on a two-dimensional display the levels of rain intensity associated with each echo shown. They also utilize colour to indicate the radial speed of the wind, the wind shear, and other information relating to severe storms.

Factors affecting Radar Performance

The performance of a radar system can be judged by the following: (1) the maximum range at which it can see a target of a specified size, (2) the accuracy of its measurement of target location in range and angle, (3) its ability to distinguish one target from another, (4) its ability to detect the desired target echo when masked by large clutter echoes, unintentional interfering signals from other "friendly" transmitters, or intentional radiation from hostile jamming (if a military radar), (5) its ability to recognize the type of target, and (6) its availability (ability to operate when needed), reliability, and maintainability.

Transmitter Power and Antenna Size

The maximum range of a radar system depends in large part on the average power of its transmitter and the physical size of its antenna. (In technical terms, this is called the power-aperture product.) There are practical limits to each. As noted before, some radar systems have an average power of roughly one megawatt. Phased-array radars about 100 feet (30 metres) in diameter are not uncommon; some are much larger. There are specialized radars with (fixed) antennas, such as some HF over-the-horizon radars and the U.S. Space Surveillance System (SPASUR), that extend more than one mile (1.6 km).

Receiver Noise

The sensitivity of a radar receiver is determined by the unavoidable noise that appears at its input. At microwave radar frequencies, the noise that limits detectability is usually generated by the receiver itself (i.e., by the random motion of electrons at the input of the receiver) rather than by external noise that enters the receiver via the antenna. A radar engineer often employs a transistor amplifier as the first stage of the receiver even though lower noise can be obtained with more sophisticated (and more complex) devices. This is an example of the application of the basic engineering principle that the "best" performance that can be obtained might not necessarily be the solution that best meets the needs of the user.

The receiver is designed to enhance the desired signals and to reduce the noise and other undesired signals that interfere with detection. A designer attempts to maximize the detectability of weak signals by using what radar engineers call a "matched filter," which is a filter that maximizes the signal-to-noise ratio at the receiver output. The matched filter has a precise mathematical formulation that depends on the shape of the input signal and the character of the receiver noise. A suitable approximation to the matched filter for the ordinary pulse radar, however, is one whose bandwidth in hertz is the reciprocal of the pulse width in seconds.

Target Size

The size of a target as "seen" by radar is not always related to the physical size of the object. The measure of the target size as observed by radar is called the radar cross section and is given in units of area (square metres). It is possible for two targets with the same physical cross-sectional area to differ considerably in radar size, or radar cross section. For example, a flat plate 1 square metre in area will produce a radar cross section of about 1,000 square metres at a frequency of 3 GHz when viewed perpendicular to the surface. A cone-sphere (an object resembling an ice-cream cone) when viewed in the direction of the cone rather than the sphere could have a radar cross section of about 0.001 square metre even though its projected area is also 1 square metre. In theory, the radar cross section has little to do with the size of the cone or the cone angle. Thus, the flat plate and the cone-sphere can have radar cross sections that differ by a million to one even though their physical projected areas are the same.

The sphere is an unusual target in that its radar cross section is the same as its physical cross-sectional area (when its circumference is large compared with the radar wavelength). That is to say, a sphere with a projected area of 1 square metre has a radar cross section of 1 square metre.

Commercial aircraft might have radar cross sections from about 10 to 100 square metres, except when viewed broadside, where the cross sections are much larger. Most air-traffic-control radars are required to detect aircraft with a radar cross section as low as 2 square metres, since some small general-aviation aircraft can be of this value. For comparison, the radar cross section of a man has been measured at microwave frequencies to be about 1 square metre. A bird can have a cross section of 0.01 to 0.001 square metre. Although this is a small value, a bird can be readily detected at ranges of several tens of kilometres by long-range radar. In general, many birds can be detected by radar, so special measures must usually be taken to ensure that their echoes do not interfere with the detection of desired targets.

The radar cross section of an aircraft and that of most other targets of practical interest fluctuate rapidly as the aspect of the target changes with respect to the radar unit. It would not be unusual for a slight change in aspect to cause the radar cross section to change by a factor of 10 to 1,000.

Clutter

Echoes from land, sea, rain, snow, hail, birds, insects, auroras, and meteors are of interest to those who observe and study the environment, but they are a nuisance to those who want to detect aircraft, ships, missiles, or other similar targets. Clutter echoes can seriously limit the capability of a radar system; thus, a significant part of radar design is devoted to minimizing the effects of clutter without reducing the echoes from desired targets. The Doppler frequency shift is the usual means by which moving targets are distinguished from the clutter of stationary objects. Detection of targets in rain is less of a problem at the lower frequencies, since the radar echo from rain decreases rapidly with decreasing frequency and the average cross section of aircraft is relatively independent of frequency in the microwave region. Because raindrops are more or less spherical (symmetrical) and aircraft are asymmetrical, the use of circular polarization can enhance the detection of aircraft in rain. With circular polarization, the electric field rotates at the radar frequency. Because of this, the electromagnetic energy reflected by the rain and the aircraft will be affected differently, which thereby makes it easier to distinguish between the two. (In fair weather most radars use linear polarization; i.e., the direction of the electric field is fixed).

Atmospheric Effects

Rain and other forms of precipitation can cause echo signals that mask the desired target echoes. There are other atmospheric phenomena that can affect radar performance as well. The decrease in density of the Earth's atmosphere with increasing altitude causes radar waves to bend as they propagate through the atmosphere. This usually increases the detection range at low angles to a slight extent. The atmosphere can form "ducts" that trap and guide radar energy around the curvature of the Earth and allow detection at ranges beyond the normal horizon. Ducting over water is more likely to occur in tropical climates than in colder regions. Ducts can sometimes extend the range of an airborne radar, but on other occasions they may cause the radar energy to be diverted and not illuminate regions below the ducts. This results in the formation of what are called radar holes in the coverage. Since it is not predictable or reliable, ducting can in some instances be more of a nuisance than a help.

Loss of radar energy due to atmospheric absorption, when propagation is through the clear atmosphere or rain, is usually small for most systems operating at microwave frequencies.

Interference

Signals from nearby radars and other transmitters can be strong enough to enter a radar receiver and produce spurious responses. Well-trained operators are not often deceived by interference, though they may find it a nuisance. Interference is not as easily ignored by automatic detection and tracking systems, however, and so some method is usually needed to recognize and remove interference pulses before they enter the automatic detector and tracker of a radar.

Electronic Countermeasures (Electronic Warfare)

The purpose of hostile electronic countermeasures (ECM) is to degrade the effectiveness of military radar deliberately. ECM can consist of (1) noise jamming that enters the receiver via the antenna and increases the noise level at the input of the receiver, (2) false target generation, or repeater jamming, by which hostile jammers introduce additional signals into the radar receiver in an attempt to confuse the receiver into thinking that they are real target echoes, (3) chaff, which is an artificial cloud consisting of a large number of tiny metallic reflecting strips that create strong echoes over a large area to mask the presence of real target echoes or to create confusion, and (4) decoys, which are small, inexpensive air vehicles or other objects designed to appear to the radar as if they are real targets. Military radars are also subject to direct attack by conventional weapons or by anti-radiation missiles (ARMs) that use radar transmissions to find the target and home in on it. A measure of the effectiveness of military radar is the large sums of money spent on electronic warfare measures, ARMs, and low-cross-section (stealth) aircraft.

Military radar engineers have developed various ways of countering hostile ECM and maintaining the ability of a radar system to perform its mission. It might be noted that a military radar system can often accomplish its mission satisfactorily even though its performance in the presence of ECM is not what it would be if such measures were absent.

Examples of Radar Systems

Airport Surveillance Radar

Airport surveillance radar systems are capable of reliably detecting and tracking aircraft at altitudes below 25,000 feet (7,620 metres) and within 40 to 60 nautical miles (75 to 110 km) of their airport. Systems of this type have been installed at more than 100 major airports throughout the United States. One such system, the ASR-9, is designed to be operable at least 99.9 percent of the time, which means that the system is down less than 10 hours per year. This high availability is attributable to reliable electronic components, a "built-in test" to search for failures, remote monitoring, and redundancy (i.e., the system has two complete channels except for the antenna; when one channel must be shut down for repair, the other continues to operate). The ASR-9 is designed to operate unattended, with no maintenance personnel at the radar site. A number of radar units can be monitored and controlled from a single location. When trouble occurs, the fault is identified and a maintenance person dispatched for repair.

Echoes from rain that mask the detection of aircraft are reduced by the use of Doppler filtering and other techniques devised to separate moving aircraft from undesired clutter. It is important, however, for air traffic controllers to recognize areas of severe weather so that they can direct aircraft safely around, rather than through, rough or hazardous conditions. The ASR-9 has a separate receiving channel that recognizes weather echoes and provides their location to air traffic controllers. Six different levels of precipitation intensity can be displayed, with or without the aircraft targets superimposed.

The ASR-9 system operates at frequencies from 2.7 to 2.9 GHz (within the S band). Its klystron transmitter has a peak power of 1.3 megawatts, a pulse width of 1 microsecond, and an antenna with a horizontal beam width of 1.4 degrees that rotates at 12.5 revolutions per minute (4.8-second rotation period).

The ASR-11 and ASR-12 are airport surveillance radars that utilize a solid-state (transistor) transmitter and long pulses rather than a klystron transmitter and short pulses.

Doppler Weather Radar

For many years radar has been used to provide information about the intensity and extent of rain and other forms of precipitation. This application of radar is well known in the United States from the familiar television weather reports of precipitation observed by the radars of the National Weather Service. A major improvement in the capability of weather radar came about when engineers developed new radars that could measure the Doppler frequency shift in addition to the magnitude of the echo signal reflected from precipitation. The Doppler frequency shift is important because it is related to the radial velocity of the precipitation blown by wind (the component of the wind moving either toward or away from the radar installation). Since tornadoes, mesocyclones (which spawn tornadoes), hurricanes, and other hazardous weather phenomena tend to rotate, measurement of the radial wind speed as a function of viewing angle will identify rotating weather patterns. (Rotation is indicated when the measurement of the Doppler frequency shift shows that the wind is coming toward the radar at one angle and away from it at a nearby angle.).

The pulse Doppler weather radars employed by the National Weather Service, which are known as Nexrad, make quantitative measurements of precipitation, warn of potential flooding or dangerous hail, provide wind speed and direction, indicate the presence of wind shear and gust fronts, track storms, predict thunderstorms, and provide other meteorological information. In addition to measuring precipitation (from the intensity of the echo signal) and radial speed (from the Doppler frequency shift), Nexrad also measures the spread in radial speed (difference between the maximum and the minimum speeds) of the precipitation particles within each radar resolution cell. The spread in radial speed is an indication of wind turbulence.

Another improvement in the weather information provided by Nexrad is the digital processing of radar data, a procedure that renders the information in a form that can be interpreted by an observer who is not necessarily a meteorologist. The computer automatically identifies severe weather effects and indicates their nature on a display viewed by the observer. High-speed communication lines integrated with the Nexrad system allow timely weather information to be transmitted for display to various users.

The Nexrad radar operates at S-band frequencies (2.7 to 3 GHz) and is equipped with a 28-foot- (8.5-metre-) diameter antenna. It takes five minutes to scan its 1 degree beamwidth through 360 degrees in azimuth and from 0 to 20 degrees in elevation. The Nexrad system can measure rainfall up to a distance of 460 km and determine its radial velocity as far as 230 km.

A serious weather hazard to aircraft in the process of landing or taking off from an airport is the downburst, or microburst. This strong downdraft causes wind shear capable of forcing aircraft to the ground. Terminal Doppler weather radar (TDWR) is the name of the type of system at or near airports that is specially designed to detect dangerous microbursts. It is similar in principle to Nexrad but is a shorter-range system since it has to observe dangerous weather phenomena only in the vicinity of an airport. It operates from 5.60 to 5.65 GHz (C band) to avoid interference with the lower frequencies of Nexrad and ASR systems.

Airborne Combat Radar

A modern combat aircraft is generally required not only to intercept hostile aircraft but also to attack surface targets on the ground or sea. The radar that serves such an aircraft must have the capabilities to perform these distinct military missions. This is not easy because each mission has different requirements. The different ranges, accuracies, and rates at which the radar data is required, the effect of the environment (land or sea clutter), and the type of target (land features or moving aircraft) call for different kinds of radar waveforms (different pulse widths and pulse repetition frequencies). In addition, an appropriate form of signal processing is required to extract the particular information needed for each military function. Radar for combat aircraft must therefore be multimode—i.e., operate with different waveforms, signal processing, and antenna scanning. It would not be unusual for an airborne combat radar to have from 8 to 10 air-to-air modes and 6 to 10 air-to-surface modes. Furthermore, the radar system might be required to assist in rendezvous with a companion combat craft or with a refueling aircraft, provide guidance of air-to-air missiles, and counter hostile electronic jamming. The problem of achieving effectiveness with these many modes is a challenge for radar designers and is made more difficult by the size and weight constraints on combat aircraft.

The AN/APG-77 radar for the U.S. Air Force F-22 stealth dual-role fighter employs what is called an active-aperture phased-array radar rather than a mechanically scanned planar-array antenna. At each radiating element of the active-aperture phased-array is an individual transmitter, receiver, phase shifter, duplexer, and control.

Ballistic Missile Defense and Satellite-surveillance Radars

The systems for detecting and tracking ballistic missiles and orbiting satellites are much larger than those for aircraft detection because the ranges are longer and the radar echoes from space targets can be smaller than echoes from aircraft. Such radars might be required to have maximum ranges of 2,000 to 3,000 nautical miles (3,700 to 5,600 km), as compared with 200 nautical miles (370 km) for a typical long-range aircraft-detection system. The average power of the transmitter for a ballistic missile defense (BMD) radar can be from several hundred kilowatts to one megawatt or more, which is about 100 times greater than the average power of radars designed for aircraft detection. Antennas for this application have dimensions on the order of tens of metres to a hundred metres or more and are electronically scanned phased-array antennas capable of steering the radar beam without moving large mechanical structures. Radar systems for long-range ballistic missile detection and satellite surveillance are commonly found at the lower frequencies (typically at frequency bands of 420–450 MHz and 1,215–1,400 MHz).

The Pave Paws radar (AN/FPS-115) is an ultrahigh-frequency (UHF; 420–450 MHz) phased-array system for detecting submarine-launched ballistic missiles. It is supposed to detect targets with a radar cross section of 10 square metres at a range of 3,000 nautical miles (5,600 km). The array antenna contains 1,792 active elements within a diameter of 72.5 feet (22 metres). Each active element is a module with its own solid-state transmitter, receiver, duplexer, and phase shifter. The total average power per antenna is about 145 kilowatts. Two antennas make up a system, with each capable of covering a sector 120 degrees in azimuth. Vertical coverage is from 3 to 85 degrees. An upgraded variant of this type of radar is used in the Ballistic Missile Early Warning System (BMEWS) network, with installations in Alaska, Greenland, and England. BMEWS is designed to provide warning of intercontinental ballistic missiles (ICBMs). Each array antenna measures about 82 feet (25 metres) across and has 2,560 active elements identical to those of the Pave Paws system. Both the BMEWS and Pave Paws radars detect and track satellites and other space objects in addition to warning of the approach of ballistic missiles.

A BMD radar has to engage one or more relatively small reentry vehicles (RVs) that carry a warhead. Ballistic missile RVs can be made to have a very low echo (low radar cross section) when illuminated by radar. They were the original low-radar-cross-section targets and appeared more than 20 years before the more highly publicized stealth aircraft became a reality in the late 1980s. Ballistic missile defense requires battle-management radars that not only detect and track a relatively small target at sufficient range to engage effectively but also must reliably distinguish the reentry vehicles that carry warheads from the many confusion targets that can be present. Confusion targets include decoys, chaff (strips of metallic foil that produce an echo similar in size to that of the reentry vehicle), exploded tank fragments, and other objects released by the attacking missile. The BMD radar must also be able to fulfill its mission in spite of hostile countermeasures and defend against ballistic missiles that can reenter at low angles (depressed trajectories). In addition, the radar must be located in a defended region and be hardened to survive either a conventional or a nuclear attack.

There are at least two basic approaches to ballistic missile defense depending on whether the RV is engaged outside the atmosphere (exoatmospheric) or within the atmosphere (endoatmospheric). Exoatmospheric engagement is attractive, since it occurs at long range and a single system can defend a large area, but it requires some reliable method to select the warhead from the many extraneous objects that can accompany the warhead. An endoatmospheric ballistic missile defense system takes advantage of the slowing down of the lighter objects (decoys, chaff, and fragments) when they reenter the atmosphere and encounter air resistance. After reentry, the heavy warhead will be separated from the accompanying lighter "junk" and thus can be engaged. A significant limitation, however, is that endoatmospheric ballistic missile defense results in a much smaller defended area.

In the 1960s there were several different systems considered for defense against ICBMs. Both the United States and the Soviet Union devised defenses, but only the Soviet Union deployed such a system, and the antiballistic missile treaty of 1972 limited it to defense of a single region (Moscow). With the increased threat from tactical ballistic missiles in the 1990s, new radar concepts were explored. One was the U.S. Army's Theater High Altitude Area Defense Ground Based Radar (THAAD GBR). This is a mobile solid-state active-aperture phased-array radar that operates within the X-band of the spectrum. A different approach to ballistic missile defense is the Israeli tactical system known as Arrow, which employs an L-band (1- to 2-GHz) active-aperture phased-array radar.

Ground-probing Radar

Radar waves are usually thought of as being reflected from the surface of the ground. However, at the lower frequencies (below several hundred megahertz), radar energy can penetrate into the ground and be reflected from buried objects. The loss in propagating in the ground is very high at these frequencies, but it is low enough to permit ranges of about 3.3 to 33 feet (1 to 10 metres) or more. This is sufficient for probing the subsurface soil in order to detect underground tunnels and utility pipes and cables, to aid in archaeological digs, and to monitor the subsurface conditions of highways and bridge roadways. The short ranges require that the radar system be able to resolve closely spaced objects, which means wide-bandwidth signals must be radiated. Normally, wide bandwidth is not available at the lower frequencies (especially when a 1-foot (30-cm) range resolution requires a 500-MHz bandwidth). However, since the energy is directed into the ground rather than radiated into space, the large frequency band needed for high resolution can be obtained without serious interference to other users of the radio spectrum.

A ground-probing radar might radiate over frequencies ranging from 5 to 500 MHz in order to obtain good penetration (which requires low frequencies) with high resolution (which requires wide bandwidth). The antenna can be placed directly on the ground. Ground-probing radar units generally are small enough to be portable.

Over-the-horizon Radar

Frequencies lower than about 100 MHz usually are not desirable for radar applications. An example where lower frequencies can provide a unique and important capability is in the shortwave or high-frequency (HF), portion of the radio band (from 3 to 30 MHz). The advantage of the HF band is that radio waves of these frequencies are refracted (bent) by the ionosphere so that the waves return to the Earth's surface at long distances beyond the horizon. This permits target detection at distances from about 500 to 2,000 nautical miles (900 to 3,700 km). Thus, an HF over-the-horizon (OTH) radar can detect aircraft at distances up to 10 times that of a ground-based microwave air-surveillance radar, whose range is limited by the curvature of the Earth. Besides detection and tracking of aircraft at long ranges, an HF OTH radar can be designed to detect ballistic missiles (particularly the disturbance caused by ballistic missiles as they travel through the ionosphere), ships, and weather effects over the ocean. Winds over the ocean generate waves on the water that can be recognized by HF OTH radar. From the Doppler frequency spectrum produced by echoes from the water waves, one can determine the direction of the waves generated by the wind and hence the direction of the wind itself. The strength of the waves (which indicates the state of the sea, or roughness) also can be ascertained. Timely information about the winds that drive waves over a wide expanse of the ocean can be obtained with HF OTH—obtainable only with great difficulty, if at all, by other means—which has proved valuable for weather prediction.

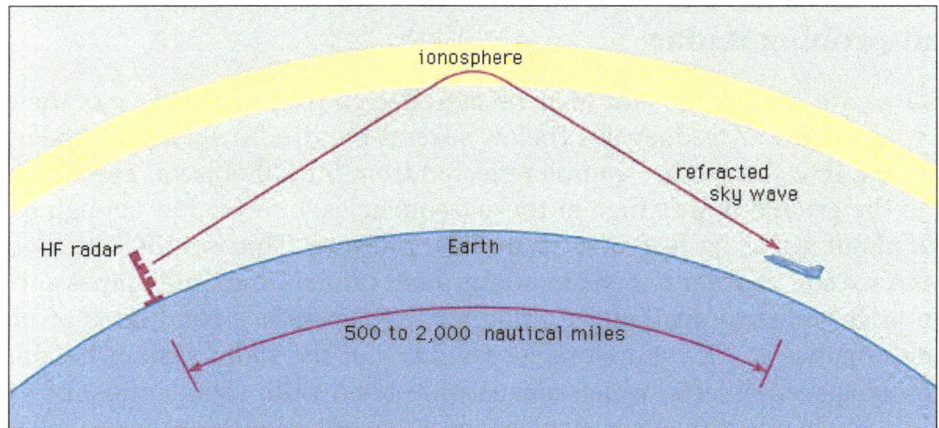

Refraction of HF radar radiation by the ionosphere.

An HF OTH radar might have an average power of about one megawatt and have phased-array antennas that sometimes extend several thousands of feet. This type of radar was originally developed for military purposes, especially for the surveillance of aircraft and ships over large expanses of water, where it is difficult for conventional microwave radars to provide coverage of large areas. For example, an important application of HF OTH is to provide wide-area surveillance of regions where illegal drug-carrying aircraft are suspected of operating. The area that can be held under surveillance by HF OTH radar is so large that it is difficult for aircraft to avoid detection by flying around or under its coverage. Furthermore, these counternarcotic radars can in many cases detect aircraft as they take off from a distant airfield and can sometimes follow them all the way to their destination. It is also possible in some cases to recognize specific aircraft types on the basis of the radar observation of the aircraft during take-off and landing. The U.S. Navy's HF OTH radars known as relocatable over-the-horizon radar (ROTHR), or AN/TPS-71, have been redirected for use in drug interdiction. Such radars, located in Virginia, Texas, and Puerto Rico, provide multiple coverage of drug-traffic regions in Central America and the northern part of South America. An ROTHR can cover a 64-degree wedge-shaped area at ranges from 500 to 1,600 nautical miles (900 to 3,000 km). Its receiving antenna is an electronically steered phased-array consisting of 372 pairs of monopole antennas. The antenna extends 1.4 nautical miles (2.5 km) in length. The transmitter operates from 5 to 28 MHz with an average power of 210 kilowatts. Each radar can provide surveillance of approximately 1.3 million square nautical miles (4.5 million square km). This is much more than 10 times the area covered by a conventional surface-based long-range microwave air-surveillance radar.

Weather Radar

Weather radar in Norman, Oklahoma with rainshaft.

Weather radar, also called weather surveillance radar (WSR) and Doppler weather radar, is a type of radar used to locate precipitation, calculate its motion, and estimate its type (rain, snow, hail etc.). Modern weather radars are mostly pulse-Doppler radars, capable of detecting the motion of rain droplets in addition to the intensity of the precipitation. Both types of data can be analyzed to determine the structure of storms and their potential to cause severe weather.

Weather (WF44) radar dish.

During World War II, radar operators discovered that weather was causing echoes on their screen, masking potential enemy targets. Techniques were developed to filter them, but scientists began to study the phenomenon. Soon after the war, surplus radars were used to detect precipitation. Since then, weather radar has evolved on its own and is now used by national weather services, research departments in universities, and in television stations' weather departments. Raw images are routinely used and specialized software can take radar data to make short term forecasts of future positions and intensities of rain, snow, hail, and other weather phenomena. Radar output is even incorporated into numerical weather prediction models to improve analyses and forecasts.

University of Oklahoma OU-PRIME C-band, polarimetric, weather radar during construction.

Working of a Weather Radar

Sending Radar Pulses

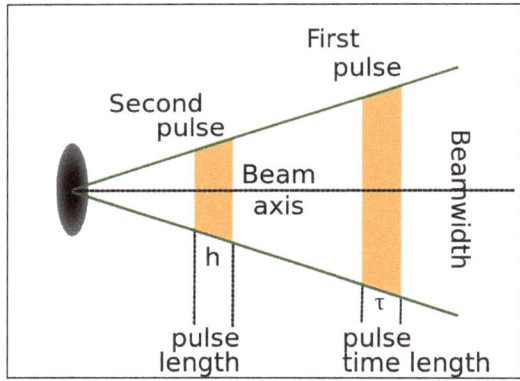

A radar beam spreads out as it moves away from the radar station, covering an increasingly large volume.

Weather radars send directional pulses of microwave radiation, on the order of a microsecond long, using a cavity magnetron or klystron tube connected by a waveguide to a parabolic antenna. The wavelengths of 1 – 10 cm are approximately ten times the diameter of the droplets or ice particles of interest, because Rayleigh scattering occurs at these frequencies. This means that part of the energy of each pulse will bounce off these small particles, back in the direction of the radar station.

Shorter wavelengths are useful for smaller particles, but the signal is more quickly attenuated. Thus 10 cm (S-band) radar is preferred but is more expensive than a 5 cm C-band system. 3 cm X-band radar is used only for short-range units, and 1 cm Ka-band weather radar is used only for research on small-particle phenomena such as drizzle and fog. W band weather radar systems have seen limited university use, but due to quicker attenuation, most data are not operational.

Radar pulses spread out as they move away from the radar station. Thus the volume of air that a radar pulse is traversing is larger for areas farther away from the station, and smaller for nearby areas, decreasing resolution at far distances. At the end of a 150 – 200 km sounding range, the volume of air scanned by a single pulse might be on the order of a cubic kilometer. This is called the *pulse volume*.

The volume of air that a given pulse takes up at any point in time may be approximated by the formula $v = hr^2\theta^2$, where v is the volume enclosed by the pulse, h is pulse width (in e.g. meters, calculated from the duration in seconds of the pulse times the speed of light), r is the distance from the radar that the pulse has already traveled (in e.g. meters), and θ is the beam width (in radians). This formula assumes the beam is symmetrically circular, "r" is much greater than "h" so "r" taken at the beginning or at the end of the pulse is almost the same, and the shape of the volume is a cone frustum of depth "h".

Listening for Return Signals

Between each pulse, the radar station serves as a receiver as it listens for return signals from particles in the air. The duration of the "listen" cycle is on the order of a millisecond, which is a thousand times longer than the pulse duration. The length of this phase is determined by the need for the microwave radiation (which travels at the speed of light) to propagate from the detector to the weather target and back again, a distance which could be several hundred kilometers. The horizontal distance from station to target is calculated simply from the amount of time that elapses from the initiation of the pulse to the detection of the return signal. The time is converted into distance by multiplying by the speed of light in air:

$$\text{Distance} = c\frac{\Delta t}{2n},$$

where $c = 299{,}792.458$ km/s is the speed of light, and $n \approx 1.0003$ is the refractive index of air.

If pulses are emitted too frequently, the returns from one pulse will be confused with the returns from previous pulses, resulting in incorrect distance calculations.

Determining Height

Assuming the Earth is round, the radar beam in vacuum would rise according to the reverse curvature of the Earth. However, the atmosphere has a refractive index that diminishes with height, due to its diminishing density. This bends the radar beam slightly toward the ground and with a standard atmosphere this is equivalent to considering that the curvature of the beam is 4/3 the actual curvature of the Earth. Depending on the elevation angle of the antenna and other considerations, the following formula may be used to calculate the target's height above ground:

$$H = \sqrt{r^2 + (k_e a_e)^2 + 2rk_e a_e \sin(\theta_e)} - k_e a_e + h_a,$$

where:

    r = distance radar–target,

    k_e = 4/3,

    a_e = Earth radius,

    a_e = elevation angle above the radar horizon,

    h_a = height of the feedhorn above ground.

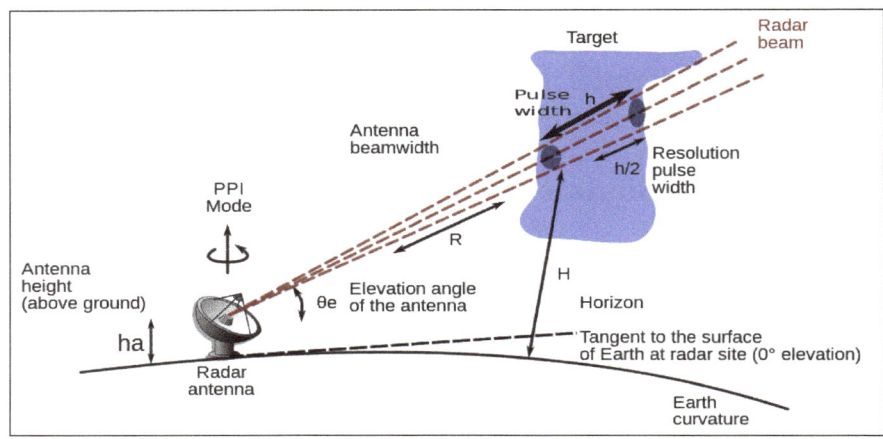

The radar beam path with height.

A weather radar network uses a series of typical angles that will be set according to the needs. After each scanning rotation, the antenna elevation is changed for the next sounding. This scenario will be repeated on many angles to scan all the volume of air around the radar within the maximum range. Usually, this scanning strategy is completed within 5 to 10 minutes to have data within 15 km above ground and 250 km distance of the radar. For instance in Canada, the 5 cm weather radars use angles ranging from 0.3 to 25 degrees. The image to the right shows the volume scanned when multiple angles are used.

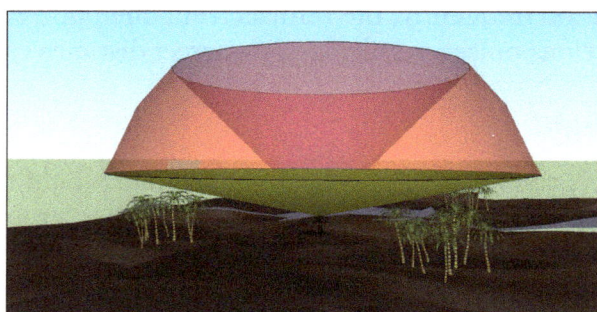

Scanned volume by using multiple elevation angles.

Due to the Earth's curvature and change of index of refraction with height, the radar cannot "see" below the height above ground of the minimal angle (shown in green) or closer to the radar than the maximal one (shown as a red cone in the center).

Calibrating Intensity of Return

Because the targets are not unique in each volume, the radar equation has to be developed beyond the basic one. Assuming a monostatic radar where $G_t = A_r \,(\text{or}\, G_r) = G$:

$$P_r = P_t \frac{G^2 \lambda^2 \sigma}{(4\pi)^3 R^4} \propto \frac{\sigma}{R^4}$$

where P_r is received power, P_t is transmitted power, G is the gain of the transmitting/receiving antenna, ë is radar wavelength, σ is the radar cross section of the target and R is the distance from transmitter to target.

In this case, we have to add the cross sections of all the targets:

$$\sigma = \bar{\sigma} = V \sum \sigma_j = V\eta$$

$$\left\{ \begin{aligned} V \quad &= \text{scanned volume} \\ &= \text{pulse length} \times \text{beam width} \\ &= \frac{c\tau}{2} \frac{\pi R^2 \theta^2}{4} \end{aligned} \right.$$

where τ is the light speed, τ is temporal duration of a pulse and θ is the beam width in radians.

In combining the two equations:

$$P_r = P_t \frac{G^2 \lambda^2}{(4\pi)^3 R^4} \frac{c\tau}{2} \frac{\pi R^2 \theta^2}{4} \eta = P_t \tau G^2 \lambda^2 \theta^2 \frac{c}{512(\pi^2)} \frac{\eta}{R^2}.$$

Which leads to:

$$P_r \propto \frac{\eta}{R^2}.$$

Notice that the return now varies inversely to R^2 instead of R^4. In order to compare the data coming from different distances from the radar, one has to normalize them with this ratio.

Data Types

Reflectivity (in Decibel or dBZ)

Return echoes from targets ("reflectivity") are analyzed for their intensities to establish the precipitation rate in the scanned volume. The wavelengths used (1–10 cm) ensure that this return is proportional to the rate because they are within the validity of

Rayleigh scattering which states that the targets must be much smaller than the wavelength of the scanning wave (by a factor of 10).

Reflectivity perceived by the radar (Z_e) varies by the sixth power of the rain droplets' diameter (D), the square of the dielectric constant (K) of the targets and the drop size distribution (e.g. N[D] of *Marshall-Palmer*) of the drops. This gives a truncated Gamma function, of the form:

$$Z_e = \int_0^{D\max} |K|^2\, N_0 e^{-\Lambda D} D^6 dD \,.$$

Precipitation rate (R), on the other hand, is equal to the number of particles, their volume and their fall speed (v[D]) as:

$$R = \int_0^{D\max} N_0 e^{-\Lambda D}\, \frac{\pi D^3}{6} v(D) dD \,.$$

So Z_e and R have similar functions that can be resolved giving a relation between the two of the form called *Z-R relation*:

Z = aR^b

Where a and b depend on the type of precipitation (snow, rain, convective or stratiform), which has different Λ, K, N_0 and v:

- As the antenna scans the atmosphere, on every angle of azimuth it obtains a certain strength of return from each type of target encountered. Reflectivity is then averaged for that target to have a better data set.

- Since variation in diameter and dielectric constant of the targets can lead to large variability in power return to the radar, reflectivity is expressed in dBZ (10 times the logarithm of the ratio of the echo to a standard 1 mm diameter drop filling the same scanned volume).

How to Read Re Lectivity on a Radar Display

NWS color scale of reflectivities.

Radar returns are usually described by colour or level. The colours in a radar image normally range from blue or green for weak returns, to red or magenta for very strong returns. The numbers in a verbal report increase with the severity of the returns. For example, the U.S. National NEXRAD radar sites use the following scale for different levels of reflectivity:

- Magenta: 65 dBZ (extremely heavy precipitation, > 16 in (410 mm) per hour, but likely hail).

- Red: 50 dBZ (heavy precipitation of 2 in (51 mm) per hour).

- Yellow: 35 dBZ (moderate precipitation of 0.25 in (6.4 mm) per hour).

- Green: 20 dBZ (light precipitation).

Strong returns (red or magenta) may indicate not only heavy rain but also thunderstorms, hail, strong winds, or tornadoes, but they need to be interpreted carefully.

Aviation Conventions

When describing weather radar returns, pilots, dispatchers, and air traffic controllers will typically refer to three return levels:

- Level 1 corresponds to a green radar return, indicating usually light precipitation and little to no turbulence, leading to a possibility of reduced visibility.

- Level 2 corresponds to a yellow radar return, indicating moderate precipitation, leading to the possibility of very low visibility, moderate turbulence and an uncomfortable ride for aircraft passengers.

- Level 3 corresponds to a red radar return, indicating heavy precipitation, leading to the possibility of thunderstorms and severe turbulence and structural damage to the aircraft.

Aircraft will try to avoid level 2 returns when possible, and will always avoid level 3 unless they are specially-designed research aircraft.

Precipitation Types

Some displays provided by commercial weather sites, like *The Weather Channel*, show precipitation types during the winter month : rain, snow, mixed precipitations (sleet and freezing rain). This is not an analysis of the radar data itself but a post-treatment done with other data sources, the primary being surface reports (METAR).

Over the area covered by radar echoes, a program assigns a precipitation type according to the surface temperature and dew point reported at the underlying weather stations. Precipitation types reported by human operated stations and certain automatic ones (AWOS) will have higher weight. Then the program does interpolations to produce an image with defined zones. These will include interpolation errors due to the calculation. Mesoscale variations of the precipitation zones will also be lost. More sophisticated programs use the numerical weather prediction output from models, such as NAM and WRF, for the precipitation types and apply it as a first guess to the radar echoes, then use the surface data for final output.

Until dual-polarization data are widely available, any precipitation types on radar images are only indirect information and must be taken with care.

Velocity

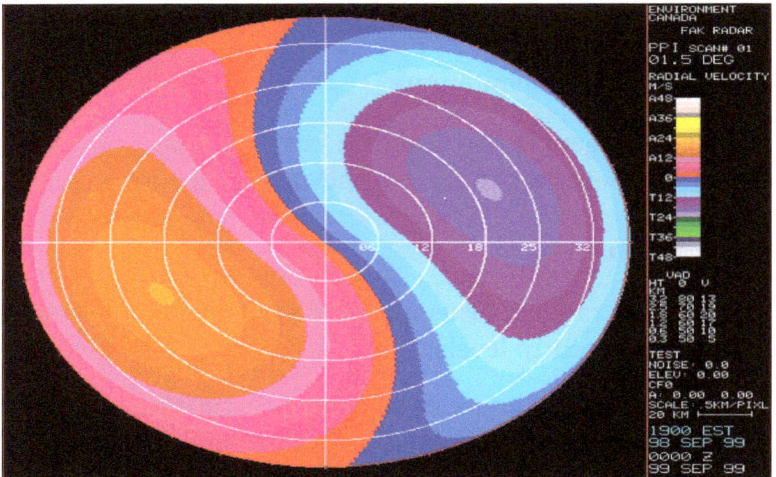

Idealized example of Doppler output. Approaching velocities are in blue and receding velocities are in red. Notice the sinusoidal variation of speed when going around the display at a particular range.

Precipitation is found in and below clouds. Light precipitation such as drops and flakes is subject to the air currents, and scanning radar can pick up the horizontal component of this motion, thus giving the possibility to estimate the wind speed and direction where precipitation is present.

A target's motion relative to the radar station causes a change in the reflected frequency of the radar pulse, due to the Doppler effect. With velocities of less than 70-metre/second for weather echos and radar wavelength of 10 cm, this amounts to a change only 0.1 ppm. This difference is too small to be noted by electronic instruments. However, as the targets move slightly between each pulse, the returned wave has a noticeable phase difference or *phase shift* from pulse to pulse.

Pulse Pair

Doppler weather radars use this phase difference (pulse pair difference) to calculate the precipitation's motion. The intensity of the successively returning pulse from the same scanned volume where targets have slightly moved is:

$$I = I_0 \sin\left(\frac{4\pi(x_0 + v\Delta t)}{\lambda}\right) = I_0 \sin\left(\Theta_0 + \Delta\Theta\right) \quad \begin{cases} x = \text{distance from radar to target} \\ \lambda = \text{radar wavelength} \\ \Delta t = \text{time between two pulses} \end{cases}$$

So $\Delta\Theta = \dfrac{4\pi v\Delta t}{\lambda}$, v = target speed = $\dfrac{\lambda\Delta\Theta}{4\pi\Delta t}$. This speed is called the radial Doppler

velocity because it gives only the radial variation of distance versus time between the radar and the target.

Doppler Dilemma

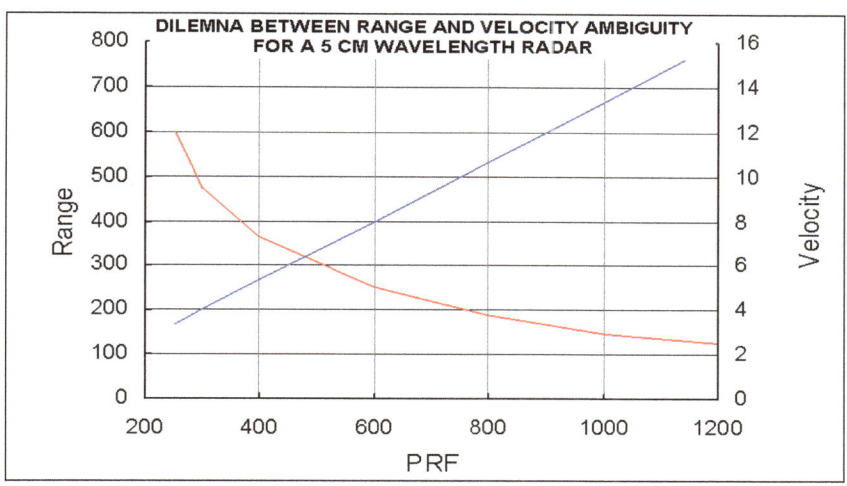

Maximum range from reflectivity (red) and unambiguous Doppler velocity range (blue) with pulse repetition frequency.

The phase between pulse pairs can vary from $-\pi$ and $+\pi$, so the unambiguous Doppler velocity range is:

$$V_{max} = \pm \frac{\lambda}{4\Delta t}.$$

This is called the Nyquist velocity. This is inversely dependent on the time between successive pulses: the smaller the interval, the larger is the unambiguous velocity range. However, we know that the maximum range from reflectivity is directly proportional to Δt:

$$X = \frac{c\Delta t}{2}.$$

The choice becomes increasing the range from reflectivity at the expense of velocity range, or increasing the latter at the expense of range from reflectivity. In general, the useful range compromise is 100–150 km for reflectivity. This means for a wavelength of 5 cm (as shown in the diagram), an unambiguous velocity range of 12.5 to 18.75 metre/second is produced (for 150 km and 100 km, respectively). For a 10 cm radar such as the NEXRAD, the unambiguous velocity range would be doubled.

Some techniques using two alternating pulse repetition frequencies (PRF) allow a greater Doppler range. The velocities noted with the first pulse rate could be equal or different with the second. For instance, if the maximum velocity with a certain rate is 10 metre/second and the one with the other rate is 15 m/s. The data coming from both will be the same up to 10 m/s, and will differ thereafter. It is then possible to find a mathematical relation between the two returns and calculate the real velocity beyond the limitation of the two PRFs.

Doppler Interpretation

In a uniform rainstorm moving eastward, a radar beam pointing west will "see" the raindrops moving toward itself, while a beam pointing east will "see" the drops moving away. When the beam scans to the north or to the south, no relative motion is noted.

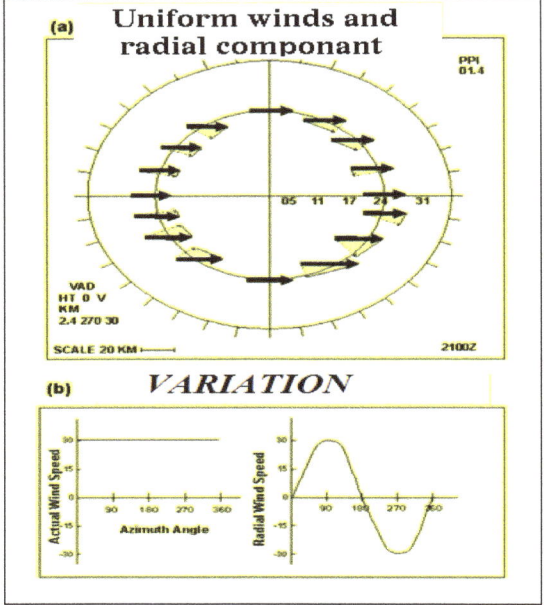

Radial component of real winds when scanning through 360 degrees.

Synoptic

In the synoptic scale interpretation, the user can extract the wind at different levels over the radar coverage region. As the beam is scanning 360 degrees around the radar, data will come from all those angles and be the radial projection of the actual wind on the individual angle. The intensity pattern formed by this scan can be represented by a cosine curve (maximum in the precipitation motion and zero in the perpendicular direction). One can then calculate the direction and the strength of the motion of particles as long as there is enough coverage on the radar screen.

However, the rain drops are falling. As the radar only sees the radial component and has a certain elevation from ground, the radial velocities are contaminated by some fraction of the falling speed. This component is negligible in small elevation angles, but must be taken into account for higher scanning angles.

Meso Scale

In the velocity data, there could be smaller zones in the radar coverage where the wind varies from the one mentioned above. For example, a thunderstorm is a mesoscale phenomenon which often includes rotations and turbulence. These may only cover few square kilometers but are visible by variations in the radial speed. Users can recognize velocity patterns in the wind associated with rotations, such as mesocyclone, convergence (outflow boundary) and divergence (downburst).

Polarization

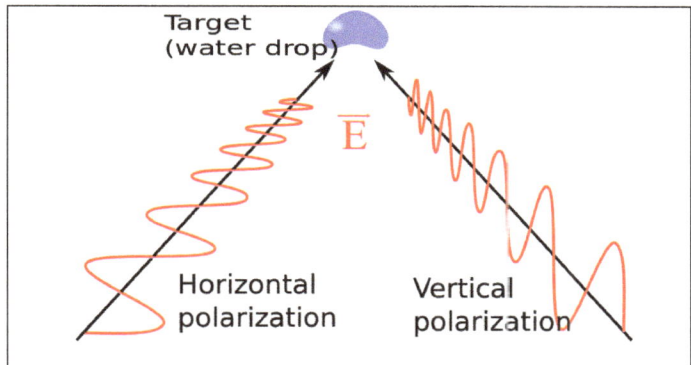

Targeting with dual-polarization will reveal the form of the droplet.

Droplets of falling liquid water tend to have a larger horizontal axis due to the drag coefficient of air while falling (water droplets). This causes the water molecule dipole to be oriented in that direction; so, radar beams are, generally, polarized horizontally in order to receive the maximal signal reflection.

If two pulses are sent simultaneously with orthogonal polarization (vertical and horizontal, Z_V and Z_H respectively), two independent sets of data will be received. These signals can be compared in several useful ways:

- Differential Reflectivity (Z_{dr}): The differential reflectivity is the ratio of the reflected vertical and horizontal power returns as Z_V/Z_H. Among other things, it is a good indicator of drop shape and drop shape is a good estimate of average drop size.

- Correlation Coefficient (ρ_{hv}): A statistical correlation between the reflected horizontal and vertical power returns. High values, near one, indicate homogeneous precipitation types, while lower values indicate regions of mixed precipitation types, such as rain and snow, or hail, or in extreme cases debris aloft, usually coinciding with a Tornado vortex signature.

- Linear Depolarization Ratio (*LDR*): This is a ratio of a vertical power return from a horizontal pulse or a horizontal power return from a vertical pulse. It can also indicate regions where there is a mixture of precipitation types.

- Differential Phase (Φ_{dp}): The differential phase is a comparison of the returned phase difference between the horizontal and vertical pulses. This change in phase is caused by the difference in the number of wave cycles (or wavelengths) along the propagation path for horizontal and vertically polarized waves. It should not be confused with the Doppler frequency shift, which is caused by the motion of the cloud and precipitation particles. Unlike the differential reflectivity, correlation coefficient and linear depolarization ratio, which are all dependent on reflected power, the differential phase is a "propagation effect." It is a very good estimator of rain rate and is not affected by attenuation. The range derivative of differential phase (specific differential phase, K_{dp}) can be used to localize areas of strong precipitation/attenuation.

With more information about particle shape, dual-polarization radars can more easily distinguish airborne debris from precipitation, making it easier to locate tornados.

With this new knowledge added to the reflectivity, velocity, and spectrum width produced by Doppler weather radars, researchers have been working on developing algorithms to differentiate precipitation types, non-meteorological targets, and to produce better rainfall accumulation estimates.In the U.S., NCAR and NSSL have been world leaders in this field.

NOAA established a test deployment for dual-polametric radar at NSSL and equipped all its 10 cm NEXRAD radars with dual-polarization, which was completed in April 2013.In 2004, ARMOR Doppler Weather Radar in Huntsville, Alabama was equipped with a SIGMET Antenna Mounted Receiver, giving Dual-Polarmetric capabilities to the operator. McGill University J. S. Marshall Radar Observatory in Montreal, Canada has converted its instrument and the data are used operationally by Environment Canada in Montreal. Another Environment Canada radar, in King City (North of Toronto), was dual-polarized in 2005; it uses a 5 cm wavelength, which experiences greater attenuation. Environment Canada is working on converting all of its radars to dual-polarization. Météo-France is planning on incorporating dual-polarizing Doppler radar in its network coverage.

Main Types of Radar Outputs

All data from radar scans are displayed according to the need of the users. Different outputs have been developed through time to reach this.

Plan Position Indicator

Since data are obtained one angle at a time, the first way of displaying them has been the Plan Position Indicator (PPI) which is only the layout of radar return on a two dimensional image. One has to remember that the data coming from different distances to the radar are at different heights above ground.

This is very important as a high rain rate seen near the radar is relatively close to what reaches the ground but what is seen from 160 km away is about 1.5 km above ground and could be far different from the amount reaching the surface. It is thus difficult to compare weather echoes at different distances from the radar.

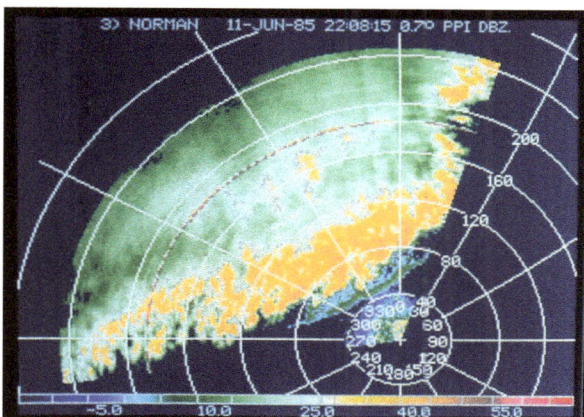

Thunderstorm line viewed in reflectivity (dBZ) on a PPI.

PPIs are afflicted with ground echoes near the radar as a supplemental problem. These can be misinterpreted as real echoes. So other products and further treatments of data have been developed to supplement such shortcomings.

Usage: Reflectivity, Doppler and polarimetric data can use PPI.

In the case of Doppler data, two points of view are possible: relative to the surface or the storm. When looking at the general motion of the rain to extract wind at different altitudes, it is better to use data relative to the radar. But when looking for rotation or wind shear under a thunderstorm, it is better to use the storm relative images that subtract the general motion of precipitation leaving the user to view the air motion as if he would be sitting on the cloud.

Constant-altitude Plan Position Indicator

To avoid some of the problems on PPIs, the constant-altitude plan position indicator (CAPPI) has been developed by Canadian researchers. It is basically a horizontal cross-section through radar data. This way, one can compare precipitation on an equal footing at difference distance from the radar and avoid ground echoes. Although data are taken at a certain height above ground, a relation can be inferred between ground stations' reports and the radar data.

CAPPIs call for a large number of angles from near the horizontal to near the vertical of the radar to have a cut that is as close as possible at all distance to the height needed. Even then, after a certain distance, there isn't any angle available and the CAPPI becomes the PPI of the lowest angle. The zigzag line on the angles diagram above shows the data used to produce 1.5 km and 4 km height CAPPIs. Notice that the section after 120 km is using the same data.

Usage: Since the CAPPI uses the closest angle to the desired height at each point from the radar, the data can originate from slightly different altitudes, as seen on the image, in different points of the radar coverage. It is therefore crucial to have a large enough number of sounding angles to minimize this height change. Furthermore, the type of data must be changing relatively gradually with height to produce an image that is not noisy.

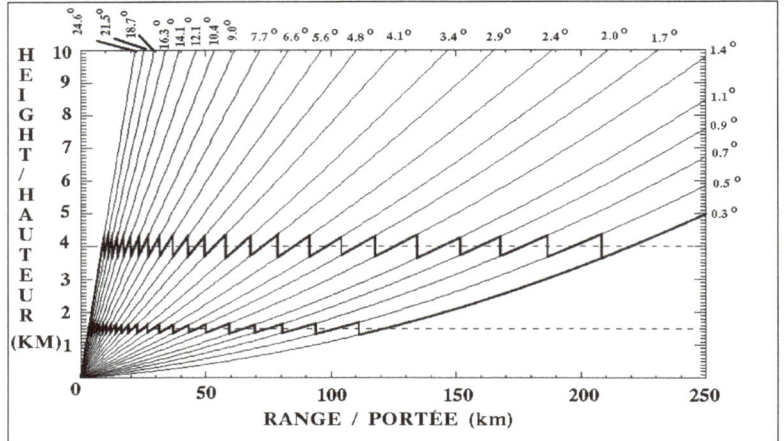

Typical angles scanned in Canada. The zigzags represent data angles used to make CAPPIs at 1.5 km and 4 km of altitude.

Reflectivity data being relatively smooth with height, CAPPIs are mostly used for displaying them. Velocity data, on the other hand, can change rapidly in direction with height and CAPPIs of them are not common. It seems that only McGill University is producing regularly Doppler CAPPIs with the 24 angles available on their radar. However, some researchers have published papers using velocity CAPPIs to study tropical cyclones and development of NEXRAD products.Finally, polarimetric data are recent and often noisy. There doesn't seem to have regular use of CAPPI for them although the *SIGMET* company offer a software capable to produce those types of images.

Vertical Composite

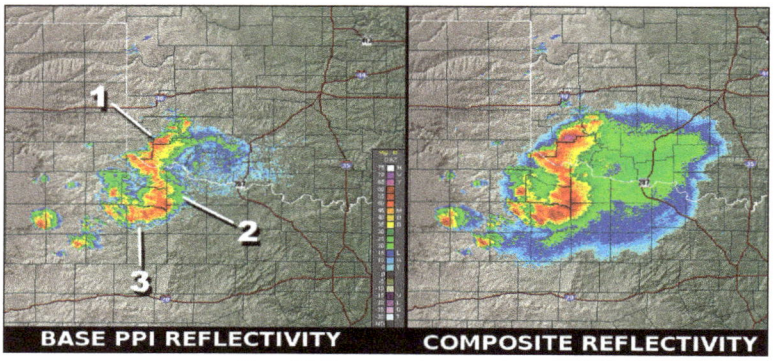

Base PPI versus Composite.

Another solution to the PPI problems is to produce images of the maximum reflectivity in a layer above ground. This solution is usually taken when the number of angles available is small or variable. The American National Weather Service is using such Composite as their scanning scheme can vary from 4 to 14 angles, according to their need, which would make very coarse CAPPIs. The Composite assures that no strong echo is missed in the layer and a treatment using Doppler velocities eliminates the ground echoes. Comparing base and composite products, one can locate virga and up-drafts zones.

Accumulations

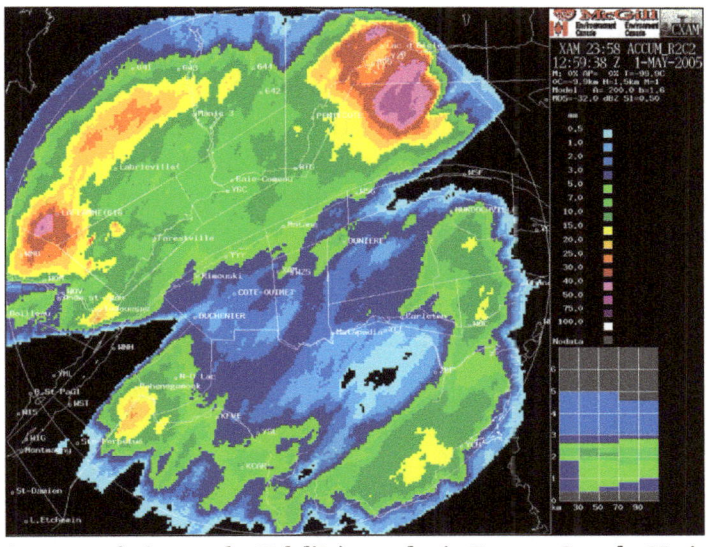

24 hours rain accumulation on the Val d'Irène radar in Eastern Canada. Notice the zones without data in the East and Southwest caused by radar beam blocking from mountains.

Another important use of radar data is the ability to assess the amount of precipitation that has fallen over large basins, to be used in hydrological calculations; such data is useful in flood control, sewer management and dam construction. The computed data from radar weather may be used in conjunction with data from ground stations.

To produce radar accumulations, we have to estimate the rain rate over a point by the average value over that point between one PPI, or CAPPI, and the next; then multiply by the time between those images. If one wants for a longer period of time, one has to add up all the accumulations from images during that time.

Echotops

Aviation is a heavy user of radar data. One map particularly important in this field is the Echotops for flight planning and avoidance of dangerous weather. Most country weather radars are scanning enough angles to have a 3D set of data over the area of coverage. It is relatively easy to estimate the maximum altitude at which precipitation is found within the volume. However, those are not the tops of clouds as they always extend above the precipitation.

Vertical Cross Sections

To know the vertical structure of clouds, in particular thunderstorms or the level of the melting layer, a vertical cross-section product of the radar data is available. This is done by displaying only the data along a line, from coordinates A to B, taken from the different angles scanned.

Range Height Indicator

When a weather radar is scanning in only one direction vertically, it obtains high resolution data along a vertical cut of the atmosphere. The output of this sounding is called a *Range Height Indicator* (RHI) which is excellent for viewing the detailed vertical structure of a storm. This is different from the vertical cross section mentioned above by the fact that the radar is making a vertical cut along specific directions and does not scan over the entire 360 degrees around the site. This kind of sounding and product is only available on research radars.

Radar Networks

Berrimah Radar in Darwin, Northern Territory Australia.

Over the past few decades, radar networks have been extended to allow the production of composite views covering large areas. For instance, many countries, including the United States, Canada and much of Europe, produce images that include all of their radars. This is not a trivial task.

In fact, such a network can consist of different types of radar with different characteristics such as beam width, wavelength and calibration. These differences have to be taken into account when matching data across the network, particularly to decide what data to use when two radars cover the same point. If one uses the stronger echo but it comes from the more distant radar, one uses returns that are from higher altitude coming from rain or snow that might evaporate before reaching the ground (virga). If one uses data from the closer radar, it might be attenuated passing through a thunderstorm. Composite images of precipitations using a network of radars are made with all those limitations in mind.

Automatic Algorithms

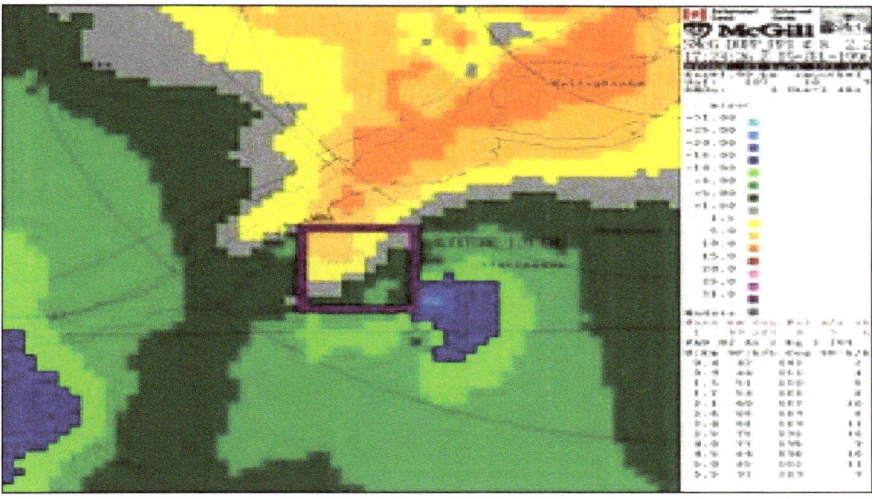

The square in this Doppler image has been automatically placed by the radar program to spot the position of a mesocyclone. Notice the inbound/outbound doublet (blue/yellow) with the zero velocity line (gray) parallel to the radial to the radar (up right). It is noteworthy to mention that the change in wind direction here occurs over less than 10 km.

To help meteorologists spot dangerous weather, mathematical algorithms have been introduced in the weather radar treatment programs. These are particularly important in analyzing the Doppler velocity data as they are more complex. The polarization data will even need more algorithms.

Main algorithms for reflectivity:

- Vertically Integrated Liquid (VIL) is an estimate of the total mass of precipitation in the clouds.

- VIL Density is VIL divided by the height of the cloud top. It is a clue to the possibility of large hail in thunderstorms.

- Potential wind gust, which can estimate the winds under a cloud (a downdraft) using the VIL and the height of the echotops (radar estimated top of the cloud) for a given storm cell.

- Hail algorithms that estimate the presence of hail and its probable size.

Main Algorithms for Doppler Velocities:

- Mesocyclone detection: it is triggered by a velocity change over a small circular area. The algorithm is searching for a "*doublet*" of inbound/outbound velocities with the zero line of velocities, between the two, along a radial line from the radar. Usually the mesocyclone detection must be found on two or more stacked progressive tilts of the beam to be significative of rotation into a thunderstorm cloud.

- TVS or Tornado Vortex Signature algorithm is essentially a mesocyclone with a large velocity threshold found through many scanning angles. This algorithm is used in NEXRAD to indicate the possibility of a tornado formation.

- Wind shear in low levels. This algorithm detects variation of wind velocities from point to point in the data and looking for a *doublet* of inbound/outbound velocities with the zero line perpendicular to the radar beam. The wind shear is associated with downdraft, (downburst and microburst), gust fronts and turbulence under thunderstorms.

- VAD Wind Profile (VWP) is a display that estimates the direction and speed of the horizontal wind at various upper levels of the atmosphere.

Animations

The animation of radar products can show the evolution of reflectivity and velocity patterns. The user can extract information on the dynamics of the meteorological phenomena, including the ability to extrapolate the motion and observe development or dissipation.

Radar Integrated Display with Geospatial Elements

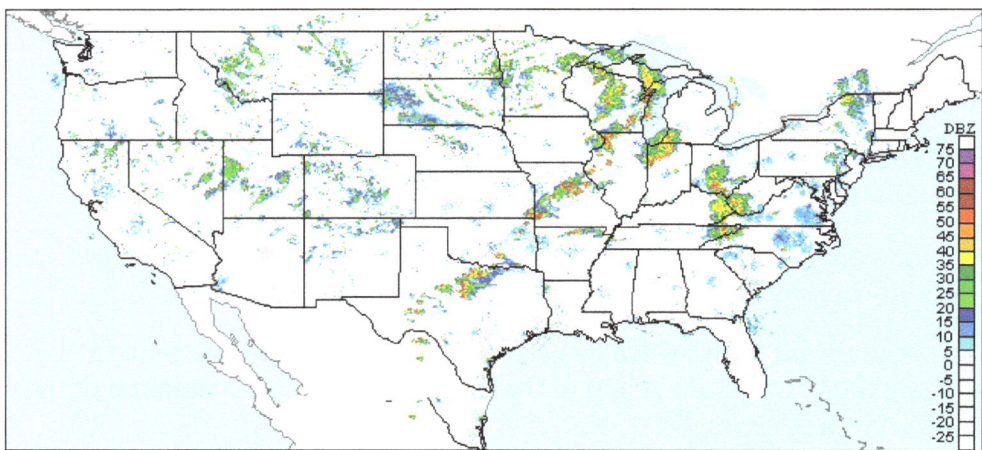

Map of the RIDGE presentation of 2011 Joplin tornado.

A new popular presentation of weather radar data in United States is via *Radar Integrated Display with Geospatial Elements* (RIDGE) in which the radar data is projected on a map with geospatial elements such as topography maps, highways, state/county boundaries and weather warnings. The projection often is flexible giving the user a choice of various geographic elements. It is frequently used in conjunction with animations of radar data over a time period.

Limitations and Artifacts

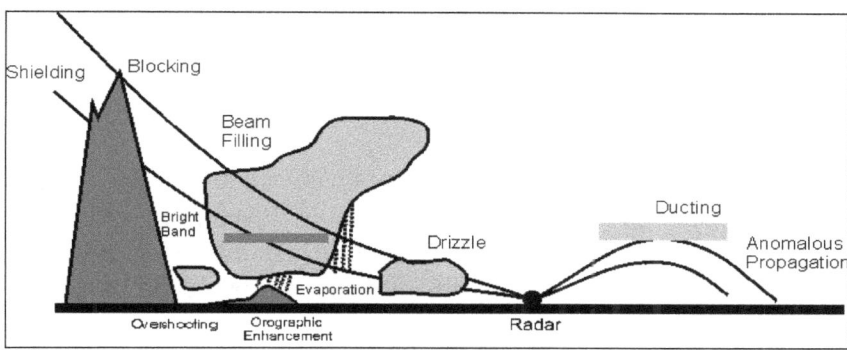

Radar data interpretation depends on many hypotheses about the atmosphere and the weather targets, including:

- International Standard Atmosphere.

- Targets small enough to obey the Rayleigh scattering, resulting in the return being proportional to the precipitation rate.

- The volume scanned by the beam is full of meteorological targets (rain, snow, etc.), all of the same variety and in a uniform concentration.

- No attenuation.

- No amplification.

- Return from side lobes of the beam are negligible.

- The beam is close to a Gaussian function curve with power decreasing to half at half the width.

- The outgoing and returning waves are similarly polarized.

- There is no return from multiple reflections.

These assumptions are not always met; one must be able to differentiate between reliable and dubious echoes.

Anomalous Propagation (Non-standard Atmosphere)

The first assumption is that the radar beam is moving through air that cools down at a certain rate with height. The position of the echoes depend heavily on this hypothesis. However, the real atmosphere can vary greatly from the norm.

Super Refraction

Temperature inversions often form near the ground, for instance by air cooling at night while remaining warm aloft. As the index of refraction of air decreases faster than normal the radar beam bends toward the ground instead of continuing upward. Eventually, it will hit the ground and be reflected back toward the radar. The processing program will then wrongly place the return echoes at the height and distance it would have been in normal conditions.

This type of false return is relatively easy to spot on a time loop if it is due to night cooling or marine inversion as one sees very strong echoes developing over an area, spreading in size laterally but not moving and varying greatly in intensity. However, inversion of temperature exists ahead of warm fronts and the abnormal propagation echoes are then mixed with real rain.

The extreme of this problem is when the inversion is very strong and shallow, the radar beam reflects many times toward the ground as it has to follow a waveguide path. This will create multiple bands of strong echoes on the radar images.

This situation can be found with inversions of temperature aloft or rapid decrease of moisture with height. In the former case, it could be difficult to notice.

Under Refraction

On the other hand, if the air is unstable and cools faster than the standard atmosphere with height, the beam ends up higher than expected. This indicates that precipitation is occurring higher than the actual height. Such an error is difficult to detect without additional temperature lapse rate data for the area.

Non-rayleigh Targets

If we want to reliably estimate the precipitation rate, the targets have to be 10 times smaller than the radar wave according to Rayleigh scattering. This is because the water molecule has to be excited by the radar wave to give a return. This is relatively true for rain or snow as 5 or 10 cm wavelength radars are usually employed.

However, for very large hydrometeors, since the wavelength is on the order of stone, the return levels off according to Mie theory. A return of more than 55 dBZ is likely to come from hail but won't vary proportionally to the size. On the other hand, very small targets such as cloud droplets are too small to be excited and do not give a recordable return on common weather radars.

Resolution and Partially Filled Scanned Volume

Radar beams have a physical dimension and data are sampled at discrete angles, not continuously, along each angle of elevation. This results in an averaging of the values of the returns for reflectivity, velocities and polarization data on the resolution volume scanned.

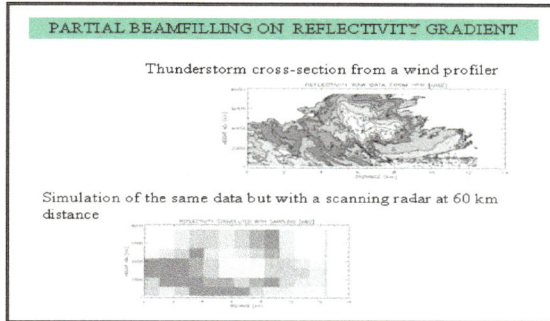

Profiler high resolution view of a thunderstorm (top) and by a weather radar (bottom).

In the figure, at the top is a view of a thunderstorm taken by a wind profiler as it was passing overhead. This is like a vertical cross section through the cloud with 150-metre vertical and 30-metre horizontal resolution. The reflectivity has large variations in a short distance. Compare this with a simulated view of what a regular weather radar would see at 60 km, in the bottom of the figure. Everything has been smoothed out. Not only the coarser resolution of the radar blur the image but the sounding incorporates area that are echo free, thus extending the thunderstorm beyond its real boundaries.

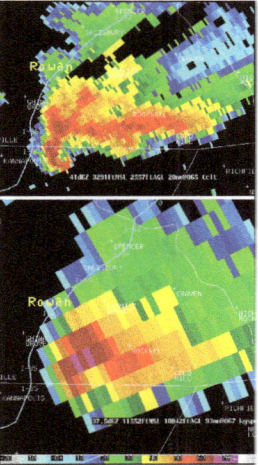

A supercell thunderstorm seen from two radars almost colocated. The top image is from a TDWR and the bottom one from a NEXRAD.

This shows how the output of weather radar is only an approximation of reality. The image compares real data from two radars almost colocated. The TDWR has about half the beamwidth of the other and one can see twice more details than with the NEXRAD.

Resolution can be improved by newer equipment but some things cannot. As mentioned previously, the volume scanned increases with distance so the possibility that the beam is only partially filled also increases. This leads to underestimation of the precipitation rate at larger distances and fools the user into thinking that rain is lighter as it moves away.

Beam Geometry

The radar beam has a distribution of energy similar to the diffraction pattern of a light passing through a slit. This is because the wave is transmitted to the parabolic antenna through a slit in the wave-guide at the focal point. Most of the energy is at the center of the beam and decreases along a curve close to a Gaussian function on each side. However, there are secondary peaks of emission that will sample the targets at off-angles from the center. Designers attempt to minimize the power transmitted by such lobes, but they cannot be completely eliminated.

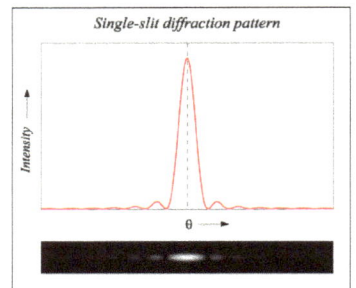

Idealized energy distribution of a radar beam (Central lobe at 0 and secondary lobes on each side).

Diffraction by a circular slit simulating the energy viewed by weather targets.

When a secondary lobe hits a reflective target such as a mountain or a strong thunderstorm, some of the energy is reflected to the radar. This energy is relatively weak but arrives at the same time that the central peak is illuminating a different azimuth. The echo is thus misplaced by the processing program. This has the effect of actually broadening the real weather echo making a smearing of weaker values on each side of it. This causes the user to overestimate the extent of the real echoes.

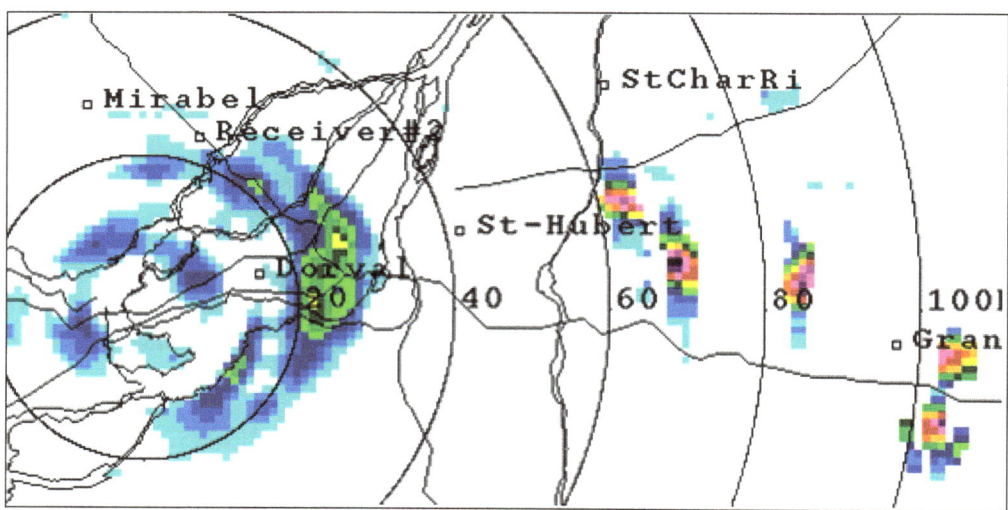

The strong echoes are returns of the central peak of the radar from a series of small hills (yellow and reds pixels). The weaker echoes on each sides of them are from secondary lobes (blue and green).

Non-weather Targets

There is more than rain and snow in the sky. Other objects can be misinterpreted as rain or snow by weather radars. Insects and arthropods are swept along by the prevailing winds, while birds follow their own course. As such, fine line patterns within weather radar imagery, associated with converging winds, are dominated by insect returns. Bird migration, which tends to occur overnight within the lowest 2000 metres of the Earth's atmosphere, contaminates wind profiles gathered by weather radar, particularly the WSR-88D, by increasing the environmental wind returns by 30–60 km/hr. Other objects within radar imagery include:

- Thin metal strips (chaff) dropped by military aircraft to fool enemies.

- Solid obstacles such as mountains, buildings, and aircraft.

- Ground and sea clutter.

- Reflections from nearby buildings ("urban spikes").

Such extraneous objects have characteristics that allow a trained eye to distinguish them. It is also possible to eliminate some of them with post-treatment of data using reflectivity, Doppler, and polarization data.

Wind Farms

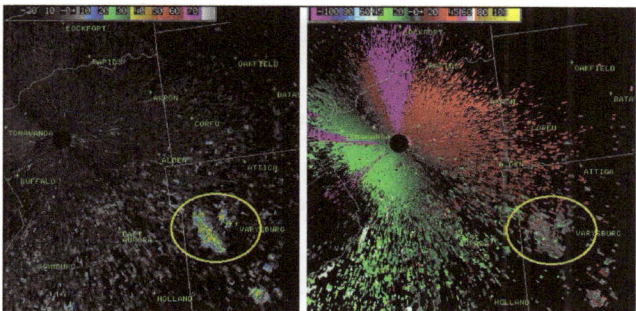

Reflectivity (left) and radial velocities (right) southeast of a NEXRAD weather radar. Echoes in circles are from a wind farm.

The rotating blades of windmills on modern wind farms can return the radar beam to the radar if they are in its path. Since the blades are moving, the echoes will have a velocity and can be mistaken for real precipitation. The closer the wind farm, the stronger the return, and the combined signal from many towers is stronger. In some conditions, the radar can even see toward and away velocities that generate false positives for the tornado vortex signature algorithm on weather radar; such an event occurred in 2009 in Dodge City, Kansas.

As with other structures that stand in the beam, attenuation of radar returns from beyond windmills may also lead to underestimation.

Attenuation

Microwaves used in weather radars can be absorbed by rain, depending on the wavelength used. For 10 cm radars, this attenuation is negligible. That is the reason why countries with high water content storms are using 10 cm wavelength, for example the US NEXRAD. The cost of a larger antenna, klystron and other related equipment is offset by this benefit.

For a 5 cm radar, absorption becomes important in heavy rain and this attenuation leads to underestimation of echoes in and beyond a strong thunderstorm. Canada and other northern countries use this less costly kind of radar as the precipitation in such areas is usually less intense. However, users must consider this characteristic when interpreting data. The images above show how a strong line of echoes seems to vanish as it moves over the radar. To compensate for this behaviour, radar sites are often chosen to somewhat overlap in coverage to give different points of view of the same storms.

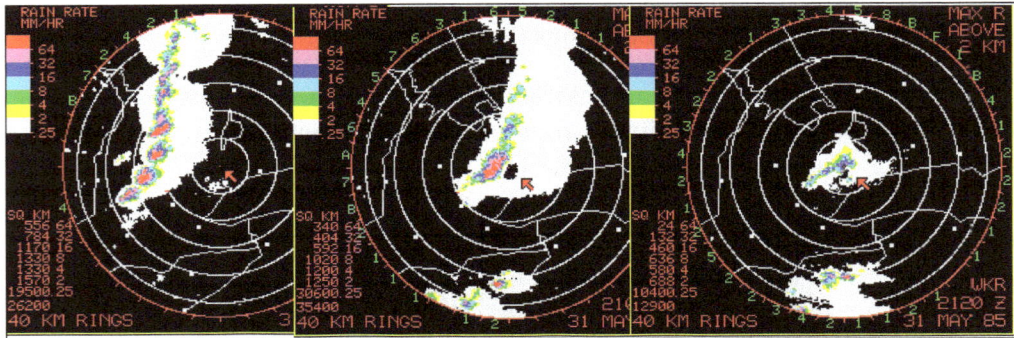

Example of strong attenuation when a line of thunderstorms moves over
(from left to right images) a 5 cm wavelength weather radar (red arrow).

Shorter wavelengths are even more attenuated and are only useful on short range radar. Many television stations in the United States have 5 cm radars to cover their audience area. Knowing their limitations and using them with the local NEXRAD can supplement the data available to a meteorologist.

Due to the spread of dual-polarization radar systems, robust and efficient approaches for the compensation of rain attenuation are currently implemented by operational weather services.

Bright Band

A radar beam's reflectivity depends on the diameter of the target and its capacity to reflect. Snowflakes are large but weakly reflective while rain drops are small but highly reflective.

When snow falls through a layer above freezing temperature, it melts into rain. Using the reflectivity equation, one can demonstrate that the returns from the snow before melting and the rain after, are not too different as the change in dielectric constant compensates for the change in size. However, during the melting process, the radar wave "sees" something akin to very large droplets as snow flakes become coated with water.

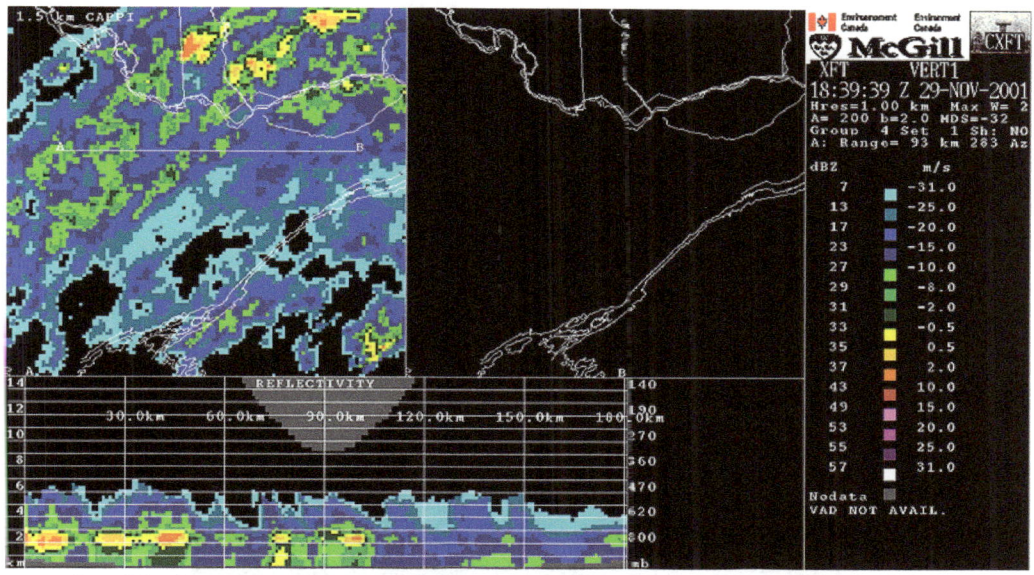

1.5 km altitude CAPPI at the top with strong contamination from the brightband (yellows). The vertical cut at the bottom shows that this strong return is only above ground.

This gives enhanced returns that can be mistaken for stronger precipitations. On a PPI, this will show up as an intense ring of precipitation at the altitude where the beam crosses the melting level while on a series of CAPPIs, only the ones near that level will have stronger echoes. A good way to confirm a bright band is to make a vertical cross section through the data, as illustrated in the picture above.

An opposite problem is that drizzle (precipitation with small water droplet diameter) tends not to show up on radar because radar returns are proportional to the sixth power of droplet diameter.

Multiple Reflections

It is assumed that the beam hits the weather targets and returns directly to the radar. In fact, there is energy reflected in all directions. Most of it is weak, and multiple reflections diminish it even further so what can eventually return to the radar from such an event is negligible. However, some situations allow a multiple-reflected radar beam to be received by the radar antenna. For instance, when the beam hits hail, the energy spread toward the wet ground will be reflected back to the hail and then to the radar. The resulting echo is weak but noticeable. Due to the extra path length it has to go through, it arrives later at the antenna and is placed further than its source.This gives a kind of triangle of false weaker reflections placed radially behind the hail.

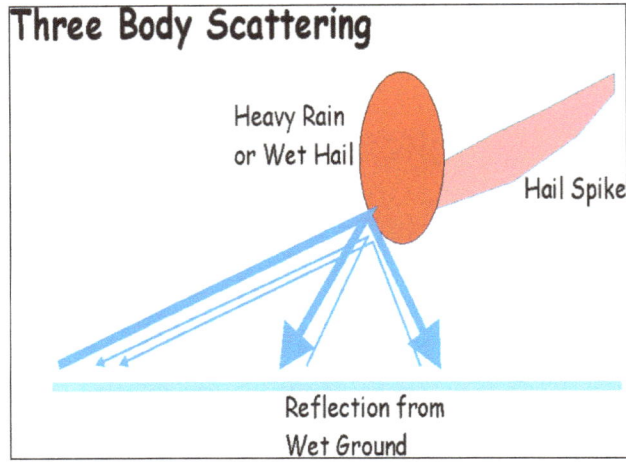

Solutions for Now and the Future

Filtering

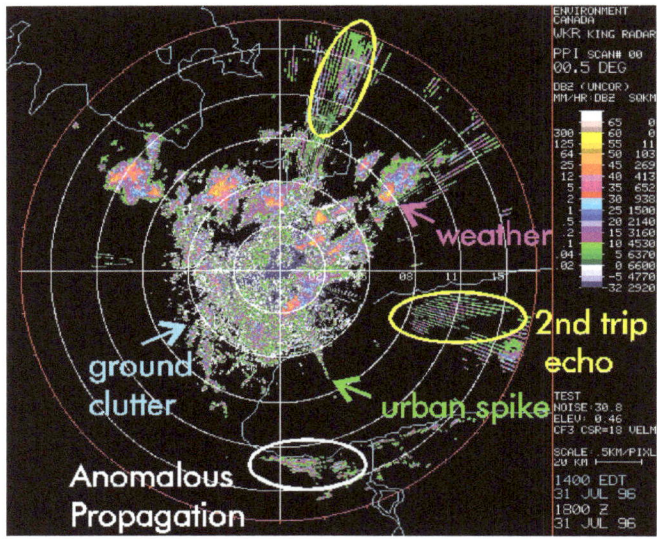

Radar image of reflectivity with many non-weather echoes.

These two images show what can be presently achieved to clean up radar data. The output on the left is made with the raw returns and it is difficult to spot the real weather. Since rain and snow clouds are usually moving, one can use the Doppler velocities to eliminate a good part of the clutter (ground echoes, reflections from buildings seen as urban spikes, anomalous propagation). The image on the right has been filtered using this property.

However, not all non-meteorological targets remain still (birds, insects, dust). Others, like the bright band, depend on the structure of the precipitation. Polarization offers a direct typing of the echoes which could be used to filter more false data or produce separate images for specialized purposes. This recent development is expected to improve the quality of radar products.

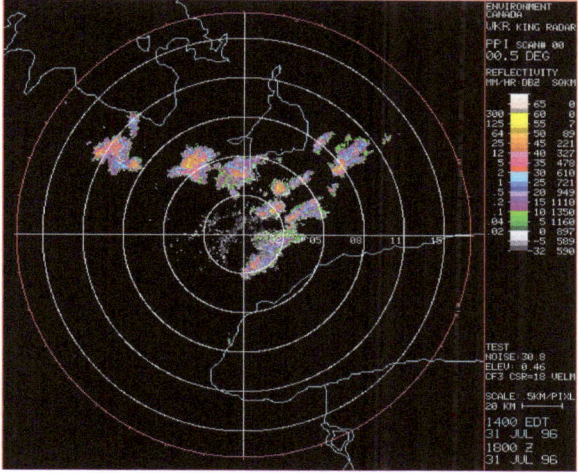

The same image but cleaned using the Doppler velocities.

Mesonet

Another question is the resolution. Radar data are an average of the scanned volume by the beam. Resolution can be improved by larger antenna or denser networks. A program by the Center for Collaborative Adaptive Sensing of the Atmosphere (CASA) aims to supplement the regular NEXRAD (a network in the United States) using many low cost X-band (3 cm) weather radar mounted on cellular telephone towers. These radars will subdivide the large area of the NEXRAD into smaller domains to look at altitudes below its lowest angle.

Phased Array Weather Radar in Norman, Oklahoma.

Using 3 cm radars, the antenna of each radar is small (about 1 meter diameter) but the resolution is similar at short distance to that of NEXRAD. The attenuation is significant due to the wavelength used but each point in the coverage area is seen by many radars, each viewing from a different direction and compensating for data lost from others.

Scanning Strategies

The number of elevation scanned and the time taken for a complete cycle depend on the weather situation. For instance, with little or no precipitation, the scheme may be limited the lowest angles and using longer impulses in order to detect wind shift near the surface. On the other hand, in violent thunderstorm situations, it is better to scan on a large number of angles in order to have a 3 dimensions view of the precipitations as often as possible. To mitigate those different demands, scanning strategies have been developed according to the type of radar, the wavelength used and the most commons weather situations in the area considered.

One example of scanning strategies is given by the US NEXRAD radar network which has evolved with time. For instance, in 2008, it added extra resolution of data,and in 2014, additional intra-cycle scanning of lowest level elevation (MESO-SAILS).

Electronic Sounding

Timeliness is also a point needing improvement. With 5 to 10 minutes time between complete scans of weather radar, much data is lost as a thunderstorm develops. A Phased-array radar is being tested at the National Severe Storms Lab in Norman, Oklahoma, to speed the data gathering. A team in Japan has also deployed a phased-array radar for 3D NowCasting at the RIKEN Advanced Institute for Computational Science (AICS).

Specialized Applications

Avionics Weather Radar

Aircraft application of radar systems include weather radar, collision avoidance, target tracking, ground proximity, and other systems. For commercial weather radar, ARINC 708 is the primary specification for weather radar systems using an airborne pulse-Doppler radar.

Global Express Weather radar with radome up.

Antennas

Unlike ground weather radar, which is set at a fixed angle, airborne weather radar is being utilized from the nose or wing of an aircraft. Not only will the aircraft be moving up, down, left, and right, but it will be rolling as well. To compensate for this, the antenna is linked and calibrated to the vertical gyroscope located on the aircraft. By doing this, the pilot is able to set a pitch or angle to the antenna that will enable the stabilizer to keep the antenna pointed in the right direction under moderate maneuvers. The small servo motors will not be able to keep up with abrupt maneuvers, but it will try. In doing this the pilot is able to adjust the radar so that it will point towards the weather system of interest. If the airplane is at a low altitude, the pilot would want to set the radar above the horizon line so that ground clutter is minimized on the display. If the airplane is at a very high altitude, the pilot will set the radar at a low or negative angle, to point the radar towards the clouds wherever they may be relative to the aircraft. If the airplane changes attitude, the stabilizer will adjust itself accordingly so that the pilot doesn't have to fly with one hand and adjust the radar with the other.

Receivers/Transmitters

There are two major systems when talking about the receiver/transmitter: the first is high-powered systems, and the second is low-powered systems; both of which operate in the X-band frequency range (8,000 – 12,500 MHz). High-powered systems operate at 10,000 – 60,000 watts. These systems consist of magnetrons that are fairly expensive (approximately $1,700) and allow for considerable noise due to irregularities with the system. Thus, these systems are highly dangerous for arcing and are not safe to be used around ground personnel. However, the alternative would be the low-powered systems. These systems operate 100 – 200 watts, and require a combination of high gain receivers, signal microprocessors, and transistors to operate as effectively as the high-powered systems. The complex microprocessors help to eliminate noise, providing a more accurate and detailed depiction of the sky. Also, since there are fewer irregularities throughout the system, the low-powered radars can be used to detect turbulence via the Doppler Effect. Since low-powered systems operate at considerable less wattage, they are safe from arcing and can be used at virtually all times.

Thunderstorm Tracking

Digital radar systems now have capabilities far beyond that of their predecessors. Digital systems now offer thunderstorm tracking surveillance. This provides users with the ability to acquire detailed information of each storm cloud being tracked. Thunderstorms are first identified by matching precipitation raw data received from the radar pulse to some sort of template preprogrammed into the system. In order for a thunderstorm to be identified, it has to meet strict definitions of intensity and shape that set it apart from any non-convective cloud. Usually, it must show signs of organization in the horizontal and continuity in the vertical: a core or a more intense center to be identified and tracked by digital radar trackers. Once the thunderstorm cell is identified, speed, distance covered, direction, and Estimated Time of Arrival (ETA) are all tracked and recorded to be utilized later.

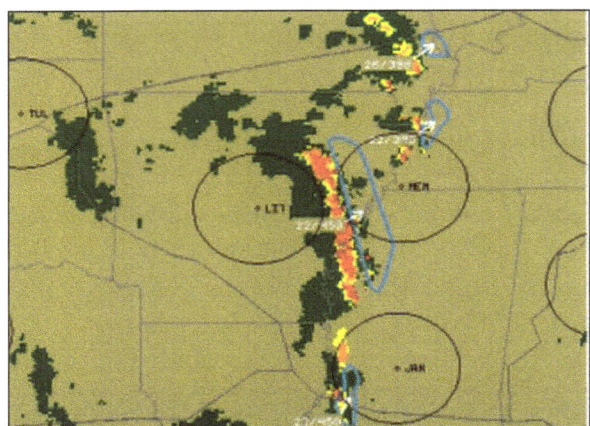

Nowcasting a line of thunderstorms from *AutoNowcaster* system.

Doppler Radar and Bird Migration

Using the Doppler weather radar is not limited to determine the location and velocity of precipitation, but it can track bird migrations as well as seen in the non-weather targets section. The radio waves sent out by the radars bounce off rain and birds alike (or even insects like butterflies). The US *National Weather Service*, for instance, have reported having the flights of birds appear on their radars as clouds and then fade away when the birds land. The U.S. National Weather Service St. Louis has even reported monarch butterflies appearing on their radars.

Different programs in North America use regular weather radars and specialized radar data to determine the paths, height of flight, and timing of migrations.This is useful information in planning for windmill farms placement and operation, to reduce bird fatalities, aviation safety and other wildlife management. In Europe, there has been similar developments and even a comprehensive forecast program for aviation safety, based on radar detection.

Doppler Radar

A Doppler radar is a specialized radar that uses the Doppler effect to produce velocity data about objects at a distance. It does this by bouncing a microwave signal off a desired target and analyzing how the object's motion has altered the frequency of the returned signal. This variation gives direct and highly accurate measurements of the radial component of a target's velocity relative to the radar. Doppler radars are used in aviation, sounding satellites, meteorology, radar guns, radiology and healthcare (fall detection and risk assessment, nursing or clinic purpose), and bistatic radar (surface-to-air missiles).

Partly because of its common use by television meteorologists in on-air weather reporting, the specific term "*Doppler Radar*" has erroneously become popularly synonymous with the type of radar used in meteorology. Most modern weather radars use the pulse-Doppler technique to examine the motion of precipitation, but it is only a part of the processing of their data. So, while these radars use a highly specialized form of *Doppler radar*, the term is much broader in its meaning and its applications.

Doppler Effect

The Doppler effect (or Doppler shift), named after Austrian physicist Christian Doppler who proposed it in 1842, is the difference between the observed frequency and the emitted frequency of a wave for an observer moving relative to the source of the waves. It is commonly heard when a vehicle sounding a siren approaches, passes and recedes from an observer. The received frequency is higher (compared to the emitted frequency) during the approach, it is identical at the instant of passing by, and it is lower during the recession. This variation of frequency also depends on the direction the wave source is moving with respect to the observer; it is maximum when the source is moving directly toward or away from the observer and diminishes with increasing angle between the direction of motion and the direction of the waves, until when the source is moving at right angles to the observer, there is no shift.

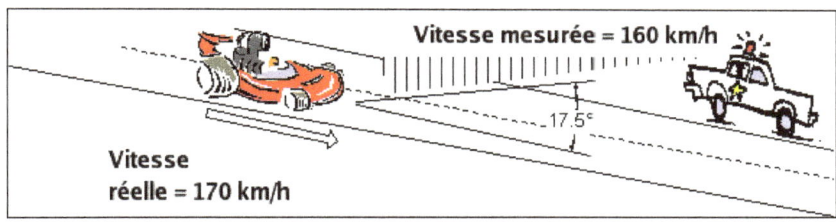

The emitted signal toward the car is reflected back with a variation of frequency that depend on the speed away/toward the radar (160 km/h). This is only a component of the real speed (170 km/h).

Imagine a baseball pitcher throwing one ball every second to a catcher (a frequency of 1 ball per second). Assuming the balls travel at a constant velocity and the pitcher is stationary, the catcher catches one ball every second. However, if the pitcher is jogging towards the catcher, the catcher catches balls more frequently because the balls are less spaced out (the frequency increases). The inverse is true if the pitcher is moving away from the catcher. The catcher catches balls less frequently because of the pitcher's backward motion (the frequency decreases). If the pitcher moves at an angle, but at the same speed, the frequency variation at which the receiver catches balls is less, as the distance between the two changes more slowly.

The frequency remains constant (whether he's throwing balls or transmitting microwaves). Since with electromagnetic radiation like microwaves or with sound, frequency is inversely proportional to wavelength, the wavelength of the waves is also affected. Thus, the relative difference in velocity between a source and an observer is what gives rise to the Doppler effect.

Frequency Variation

The formula for radar Doppler shift is the same as that for reflection of light by a moving mirror. There is no need to invoke Einstein's theory of special relativity, because all observations are made in the same frame of reference. The result derived with c as the speed of light and v as the target velocity gives the shifted frequency (f_r) as a function of the original frequency (f_t):

$$f_r = f_t \left(\frac{1 + v/c}{1 - v/c} \right),$$

which simplifies to:

$$f_r = f_t \left(\frac{c + v}{c - v} \right).$$

The "beat frequency", (Doppler frequency) (f_d), is thus:

$$f_d = f_r - f_t = 2v \frac{f_t}{(c - v)}.$$

Since for most practical applications of radar, $v \ll c$, so $(c - v) \rightarrow c$. We can then write:

$$f_d \approx 2v \frac{f_t}{c}.$$

Technology

There are four ways of producing the Doppler effect. Radars may be:

- Coherent pulsed (CP),
- Pulse-Doppler radar,
- Continuous wave (CW),
- Frequency modulation (FM).

Doppler allows the use of narrow band receiver filters that reduce or eliminate signals from slow moving and stationary objects. This effectively eliminates false signals produced by trees, clouds, insects, birds, wind, and other environmental influences. Cheap hand held Doppler radar may produce erroneous measurements.

CW Doppler radar only provides a velocity output as the received signal from the target is compared in frequency with the original signal. Early Doppler radars included CW, but these quickly led to the development of frequency modulated continuous wave (FMCW) radar, which sweeps the transmitter frequency to encode and determine range.

U.S. Army soldier using a radar gun, an application of
Doppler radar, to catch speeding violators.

With the advent of digital techniques, Pulse-Doppler radars (PD) became light enough for aircraft use, and Doppler processors for coherent pulse radars became more common. That provides Look-down/shoot-down capability. The advantage of combining Doppler processing with pulse radars is to provide accurate velocity information. This velocity is called range-rate. It describes the rate that a target moves toward or away from the radar. A target with no range-rate reflects a frequency near the transmitter frequency and cannot be detected. The classic zero doppler target is one which is on a heading that is tangential to the radar antenna beam. Basically, any target that is heading 90 degrees in relation to the antenna beam cannot be detected by its velocity.

Ultra-wideband waveforms have been investigated by the U.S. Army Research Laboratory (ARL) as a potential approach to Doppler processing due to its low average power, high resolution, and object-penetrating ability. While investigating the feasibility of whether UWB radar technology can incorporate Doppler processing to estimate the velocity of a moving target when the platform is stationary, a 2013 ARL report highlighted issues related to target range migration. However, researchers have suggested that these issues can be alleviated if the correct matched filter is used.

In military airborne applications, the Doppler effect has 2 main advantages. Firstly, the radar is more robust against counter-measure. Return signals from weather, terrain, and countermeasures like chaff are filtered out before detection, which reduces computer and operator loading in hostile environments. Secondly, against a low altitude target, filtering on the radial speed is a very effective way to eliminate the ground clutter that always has a null speed. Low-flying military plane with countermeasure alert for hostile radar track acquisition can turn perpendicular to the hostile radar to nullify its Doppler frequency, which usually breaks the lock and drives the radar off by hiding against the ground return which is much larger.

Synthetic-aperture Radar

Synthetic-aperture radar (SAR) is a form of radar that is used to create two-dimensional images or three-dimensional reconstructions of objects, such as landscapes. SAR uses the motion of the radar antenna over a target region to provide finer spatial resolution than conventional beam-scanning radars. SAR is typically mounted on a moving platform, such as an aircraft or spacecraft, and has its origins in an advanced form of side looking airborne radar (SLAR). The distance the SAR device travels over a target in the time taken for the radar pulses to return to the antenna creates the large *synthetic* antenna aperture (the *size* of the antenna). Typically, the larger the aperture, the higher the image resolution will be, regardless of whether the aperture is physical (a large antenna) or synthetic (a moving antenna) – this allows SAR to create high-resolution images with comparatively small physical antennas. Additionally, SAR has the property of having larger apertures for more distant objects, allowing consistent spatial resolution over a range of viewing distances.

This radar image acquired by the SIR-C/X-SAR radar on board the Space Shuttle Endeavour shows the Teide volcano. The city of Santa Cruz de Tenerife is visible as the purple and white area on the lower right edge of the island. Lava flows at the summit crater appear in shades of green and brown, while vegetation zones appear as areas of purple, green and yellow on the volcano's flanks.

To create a SAR image, successive pulses of radio waves are transmitted to "illuminate" a target scene, and the echo of each pulse is received and recorded. The pulses are transmitted and the echoes received using a single beam-forming antenna, with wavelengths of a meter down to several millimeters. As the SAR device on board the aircraft or spacecraft moves, the antenna location relative to the target changes with time. Signal processing of the successive recorded radar echoes allows the combining of the recordings from these multiple antenna positions. This process forms the *synthetic antenna aperture* and allows the creation of higher-resolution images than would otherwise be possible with a given physical antenna.

As of 2010, airborne systems provide resolutions of about 10 cm, ultra-wideband systems provide resolutions of a few millimeters, and experimental terahertz SAR has provided sub-millimeter resolution in the laboratory.

Motivation and Applications

SAR is capable of high-resolution remote sensing, independent of flight altitude, and independent of weather, as SAR can select frequencies to avoid weather caused signal attenuation. SAR has day and night imaging capability as illumination is provided by the SAR.

SAR images have wide application in remote sensing and mapping of surfaces of the Earth and other planets. Applications of SAR include topography, oceanography, glaciology, geology (for example, terrain discrimination and subsurface imaging), and forestry, including forest height, biomass, deforestation. Volcano and earthquake monitoring use differential interferometry. SAR can also be applied for monitoring civil infrastructure stability such as bridges. SAR is useful in environment monitoring such as oil spills, flooding, urban growth, global change and military surveillance, including strategic policy and tactical assessment. SAR can be implemented as inverse SAR by observing a moving target over a substantial time with a stationary antenna.

Basic Principle

The surface of Venus, as imaged by the Magellan probe using SAR.

A *synthetic-aperture radar* is an imaging radar mounted on a moving platform. Electromagnetic waves are transmitted sequentially, the echoes are collected and the system electronics digitizes and stores the data for subsequent processing. As transmission and reception occur at different times, they map to different positions. The well ordered combination of the received signals builds a virtual aperture that is much longer than the physical antenna width. That is the source of the term "synthetic aperture," giving it the property of an imaging radar. The range direction is parallel to the flight track and perpendicular to the azimuth direction, which is also known as the *along-track* direction because it is in line with the position of the object within the antenna's field of view.

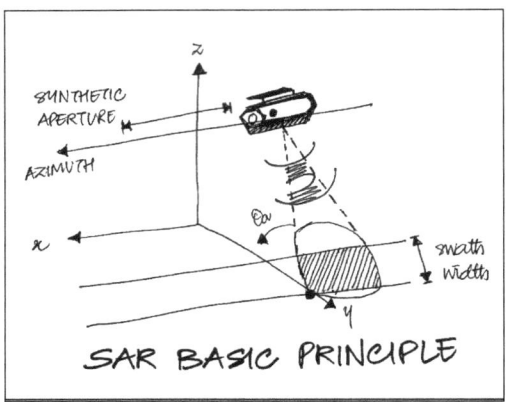

Basic principle.

The 3D processing is done in two stages. The azimuth and range direction are focused for the generation of 2D (azimuth-range) high-resolution images, after which a digital elevation model (DEM) is used to measure the phase differences between complex images, which is determined from different look angles to recover the height information. This height information, along with the azimuth-range coordinates provided by 2-D SAR focusing, gives the third dimension, which is the elevation. The first step requires only standard processing algorithms, for the second step, additional pre-processing such as image co-registration and phase calibration is used.

In addition, multiple baselines can be used to extend 3D imaging to the *time dimension*. 4D and multi-D SAR imaging allows imaging of complex scenarios, such as urban areas, and has improved performance with respect to classical interferometric techniques such as persistent scatterer interferometry (PSI).

Algorithm

The SAR algorithm, as given here, generally applies to phased arrays.

A three-dimensional array (a volume) of scene elements is defined, which will represent the volume of space within which targets exist. Each element of the array is a cubical voxel representing the probability (a "density") of a reflective surface being at that location in space.

Initially, the SAR algorithm gives each voxel a density of zero.

Then for each captured waveform, the entire volume is iterated. For a given waveform and voxel, the distance from the position represented by that voxel to the antennas used to capture that waveform is calculated. That distance represents a time delay into the waveform. The sample value at that position in the waveform is then added to the voxel's density value. This represents a possible echo from a target at that position. Note there are several optional approaches here, depending on the precision of the waveform timing, among other things. For example, if phase cannot be accurately determined, only the envelope magnitude (with the help of a Hilbert transform) of the waveform sample might be added to the voxel. If waveform polarization and phase are known and are accurate enough, then these values might be added to a more complex voxel that holds such measurements separately.

After all waveforms have been iterated over all voxels, the basic SAR processing is complete.

What remains, in the simplest approach, is to decide what voxel density value represents a solid object. Voxels whose density is below that threshold are ignored. Note the threshold level chosen must be higher than the peak energy of any single wave, otherwise that wave peak would appear as a sphere (or ellipse, in the case of multi static operation) of false "density" across the entire volume. Thus to detect a point on a target, there must be at least two different antenna echoes from that point. Consequently, there is a need for large numbers of antenna positions to properly characterize a target.

The voxels that passed the threshold criteria are visualized in 2D or 3D. Optionally, added visual quality can sometimes be had by use of a surface detection algorithm like marching cubes.

Existing Spectral Estimation Approaches

Synthetic-aperture radar determines the 3D reflectivity from measured SAR data. It is basically a spectrum estimation, because for a specific cell of an image, the complex-value SAR measurements of the SAR image stack are a sampled version of the Fourier transform of reflectivity in elevation direction, but the Fourier transform is irregular.Thus the spectral estimation techniques are used to improve the resolution and reduce speckle compared to the results of conventional Fourier transform SAR imaging techniques.

Non-parametric Methods

FFT

FFT (i.e., periodogram or matched filter) is one such method, which is used in majority of the spectral estimation algorithms, and there are many fast algorithms for computing the multidimensional discrete Fourier transform. Computational *Kronecker-core array algebra* is a popular algorithm used as new variant of FFT algorithms for the processing in multidimensional synthetic-aperture radar (SAR) systems. This algorithm uses a study of theoretical properties of input/output data indexing sets and groups of permutations.

A branch of finite multi-dimensional linear algebra is used to identify similarities and differences among various FFT algorithm variants and to create new variants. Each multidimensional DFT computation is expressed in matrix form. The multidimensional

DFT matrix, in turn, is disintegrated into a set of factors, called functional primitives, which are individually identified with an underlying software/hardware computational design.

The FFT implementation is essentially a realization of the mapping of the mathematical framework through generation of the variants and executing matrix operations. The performance of this implementation may vary from machine to machine, and the objective is to identify on which machine it performs best.

Advantages:

- Additive group-theoretic properties of multidimensional input/output indexing sets are used for the mathematical formulations, therefore, it is easier to identify mapping between computing structures and mathematical expressions, thus, better than conventional methods.

- The language of CKA algebra helps the application developer in understanding which are the more computational efficient FFT variants thus reducing the computational effort and improve their implementation time.

Disadvantages:

- FFT cannot separate sinusoids close in frequency. If the periodicity of the data does not match FFT, edge effects are seen.

Capon Method

The Capon spectral method, also called the minimum-variance method, is a multidimensional array-processing technique. It is a nonparametric covariance-based method, which uses an adaptive matched-filterbank approach and follows two main steps:

- Passing the data through a 2D bandpass filter with varying center frequencies (ω_1, ω_2).

- Estimating the power at (ω_1, ω_2) for all $\omega_1 \in [0, 2\pi), \omega_2 \in [0, 2\pi)$ of interest from the filtered data.

The adaptive Capon bandpass filter is designed to minimize the power of the filter output, as well as pass the frequencies (ω_1, ω_2) without any attenuation, i.e., to satisfy, for each (ω_1, ω_2),

$$\min h_{\omega_1,\omega_2} R h_{\omega_1,\omega_2} \text{ subject to } h^*_{\omega_1,\omega_2} a_{\omega_1,\omega_2} = 1,$$

where R is the covariance matrix, $h^*_{\omega_1,\omega_2}$ is the complex conjugate transpose of the impulse response of the FIR filter, a_{ω_1,ω_2} is the 2D Fourier vector, defined as $a_{\omega_1,\omega_2} \triangleq a_{\omega_1} \otimes a_{\omega_2}$, $\otimes$ denotes Kronecker product.

Therefore, it passes a 2D sinusoid at a given frequency without distortion while minimizing the variance of the noise of the resulting image. The purpose is to compute the spectral estimate efficiently.

Spectral estimate is given as:

$$S_{\omega_1,\omega_2} = \frac{1}{a^*_{\omega_1,\omega_2} R^{-1} a_{\omega_1,\omega_2}},$$

where R is the covariance matrix, and $a^*_{\omega_1,\omega_2}$ is the 2D complex-conjugate transpose of the Fourier vector. The computation of this equation over all frequencies is time consuming. It is seen that the forward–backward Capon estimator yields better estimation than the forward-only classical capon approach. The main reason behind this is that while the forward–backward Capon uses both the forward and backward data vectors to obtain the estimate of the covariance matrix, the forward-only Capon uses only the forward data vectors to estimate the covariance matrix.

Advantages:

- Capon can yield more accurate spectral estimates with much lower sidelobes and narrower spectral peaks than the fast Fourier transform (FFT) method.

- Capon method can provide much better resolution.

Disadvantages:

- Implementation requires computation of two intensive task: inversion of the covariance matrix R and also multiply it with the a_{ω_1,ω_2} matrix, which has to be done for each point (ω_1,ω_2).

APES Method

The APES (amplitude and phase estimation) method is also a matched-filter-bank method, which assumes that the phase history data is a sum of 2D sinusoids in noise.

APES spectral estimator has 2-step filtering interpretation:

- Passing data through a bank of FIR bandpass filters with varying center frequency ω.

- Obtaining the spectrum estimate for $\omega \in [0, 2\pi)$ from the filtered data.

Empirically, the APES method results in wider spectral peaks than the Capon method, but more accurate spectral estimates for amplitude in SAR. In the Capon method, although the spectral peaks are narrower than the APES, the sidelobes are higher than that for the APES. As a result, the estimate for the amplitude is expected to be less accurate for the Capon method than for the APES method. The APES method requires about 1.5 times more computation than the Capon method.

Advantages:

- Filtering reduces the number of available samples, but when it is designed tactically, the increase in signal-to-noise ratio (SNR) in the filtered data will compensate this reduction, and the amplitude of a sinusoidal component with frequency   can be estimated more accurately from the filtered data than from the original signal.

Disadvantages:

- The auto covariance matrix is much larger in 2D than in 1D, therefore it is limited by memory available.

SAMV Method

SAMV method is a parameter-free sparse signal reconstruction based algorithm. It achieves superresolution and robust to highly correlated signals. The name emphasizes its basis on the asymptotically minimum variance (AMV) criterion. It is a powerful tool for the recovery of both the amplitude and frequency characteristics of multiple highly correlated sources in challenging environment (e.g., limited number of snapshots, low signal-to-noise ratio. Applications include synthetic-aperture radar imaging and various source localization.

Advantages:

- SAMV method is capable of achieving resolution higher than some established parametric methods, e.g., MUSIC, especially with highly correlated signals.

Disadvantages:

- Computation complexity of the SAMV method is higher due to its iterative procedure.

Parametric Subspace Decomposition Methods

Eigenvector Method

This subspace decomposition method separates the eigenvectors of the autocovariance matrix into those corresponding to signals and to clutter. The amplitude of the image at a point (ω_x, ω_y) is given by:

$$\hat{\phi}_{EV}\left(\omega_x, \omega_y\right) = \frac{1}{W^H\left(\omega_x, \omega_y\right)\left(\sum_{\text{clutter}} \frac{1}{\lambda_i} \underline{v}_i \underline{v}_i^H\right) W\left(\omega_x, \omega_y\right)}$$

where $\hat{\phi}$ is the amplitude of the image at a point $\hat{\phi}$, $\underline{v}_i$ is the coherency matrix and $\underline{v}_i^H$ is the Hermitian of the coherency matrix,—is the inverse of the eigenvalues of the clutter

subspace, $W\left(\omega_x,\omega_y\right)$ are vectors defined as:

$$W\left(\omega_x,\omega_y\right)=\left[1\exp\left(-j\omega_x\right)\ldots\exp\left(-j(M-1)\omega_x\right)\right]\otimes\left[1\exp\left(-j\omega_y\right)\ldots\exp\left(-j(M-1)\omega_y\right)\right]$$

where $\otimes$ denotes the Kronecker product of the two vectors.

Advantages:

- Shows features of image more accurately.

Disadvantages:

- High computational complexity.

MUSIC Method

MUSIC detects frequencies in a signal by performing an eigen decomposition on the co-variance matrix of a data vector of the samples obtained from the samples of the received signal. When all of the eigenvectors are included in the clutter subspace (model order = 0) the EV method becomes identical to the Capon method. Thus the determination of model order is critical to operation of the EV method. The eigenvalue of the R matrix decides whether its corresponding eigenvector corresponds to the clutter or to the signal subspace.

The MUSIC method is considered to be a poor performer in SAR applications. This method uses a constant instead of the clutter subspace.

In this method, the denominator is equated to zero when a sinusoidal signal corresponding to a point in the SAR image is in alignment to one of the signal subspace eigenvectors which is the peak in image estimate. Thus this method does not accurately represent the scattering intensity at each point, but show the particular points of the image.

Advantages:

- MUSIC whitens or equalizes, the clutter eigenvalues.

Disadvantages:

- Resolution loss due to the averaging operation.

Backprojection Algorithm

Backprojection Algorithm has two methods: *Time-domain Backprojection* and *Frequency-domain Backprojection*. The time-domain Backprojection has more advantages over frequency-domain and thus, is more preferred. The time-domain Backprojection forms images or spectrums by matching the data acquired from the radar and as per what it expects to receive. It can be considered as an ideal matched-filter for synthetic-aperture radar. There is no need of having a different motion compensation step due to its quality of handling non-ideal motion/sampling. It can also be used for various imaging geometries.

Advantages:

- It is invariant to the imaging mode: Which means, that it uses the same algorithm irrespective of the imaging mode present, whereas, frequency domain methods require changes depending on the mode and geometry.

- Ambiguous azimuth aliasing usually occurs when the Nyquist spatial sampling requirements are exceeded by frequencies. Unambiguous aliasing occurs in squinted geometries where the signal bandwidth does not exceed the sampling limits, but has undergone "spectral wrapping." Backprojection Algorithm does not get affected by any such kind of aliasing effects.

- It matches the space/time filter: Uses the information about the imaging geometry, to produce a pixel-by-pixel varying matched filter to approximate the expected return signal. This usually yields antenna gain compensation.

- With reference to the previous advantage, the back projection algorithm compensates for the motion. This becomes an advantage at areas having low altitudes.

Disadvantages:

- The computational expense is more for Backprojection algorithm as compared to other frequency domain methods.

- It requires very precise knowledge of imaging geometry.

Application: Geosynchronous Orbit Synthetic-Aperture Radar (GEO-SAR)

In GEO-SAR, to focus specially on the relative moving track, the backprojection algorithm works very well. It uses the concept of Azimuth Processing in the time domain. For the satellite-ground geometry, GEO-SAR plays a significant role.

The procedure of this concept is elaborated as follows:

- The raw data acquired is segmented or drawn into sub-apertures for simplification of speedy conduction of procedure.

- The range of the data is then compressed, using the concept of "Matched Filtering" for every segment/sub-aperture created. It is given by: $s(t,\tau) = \exp\left(-j \cdot \dfrac{4\pi}{\lambda} \cdot R(t)\right) \cdot sinc\left(\tau - \dfrac{2}{c} \cdot R(t)\right)$ where τ is the range time, t is the azimuthal time, λ is the wavelength, c is the speed of light.

- Accuracy in the "Range Migration Curve" is achieved by range interpolation.

- The pixel locations of the ground in the image is dependent on the satellite–ground geometry model. Grid-division is now done as per the azimuth time.

- Calculations for the "slant range" (range between the antenna's phase center and the point on the ground) are done for every azimuth time using coordinate transformations.

- Azimuth Compression is done after the previous step.

- Step 5 and 6 are repeated for every pixel, to cover every pixel, and conduct the procedure on every sub-aperture.

- Lastly, all the sub-apertures of the image created throughout, are superimposed onto each other and the ultimate HD image is generated.

Comparison between the Algorithms

Capon and APES can yield more accurate spectral estimates with much lower sidelobes and more narrow spectral peaks than the fast Fourier transform (FFT) method, which is also a special case of the FIR filtering approaches. It is seen that although the APES algorithm gives slightly wider spectral peaks than the Capon method, the former yields more accurate overall spectral estimates than the latter and the FFT method.

FFT method is fast and simple but have larger sidelobes. Capon has high resolution but high computational complexity. EV also has high resolution and high computational complexity. APES has higher resolution, faster than capon and EV but high computational complexity.

MUSIC method is not generally suitable for SAR imaging, as whitening the clutter eigenvalues destroys the spatial inhomogeneities associated with terrain clutter or other diffuse scattering in SAR imagery. But it offers higher frequency resolution in the resulting power spectral density (PSD) than the fast Fourier transform (FFT)-based methods.

The backprojection algorithm is computationally expensive. It is specifically attractive for sensors that are wideband, wide-angle, and have long coherent apertures with substantial off-track motion.

More Complex Operation

The basic design of a synthetic-aperture radar system can be enhanced to collect more information. Most of these methods use the same basic principle of combining many pulses to form a synthetic aperture, but may involve additional antennas or significant additional processing.

Multistatic Operation

SAR requires that echo captures be taken at multiple antenna positions. The more captures taken (at different antenna locations) the more reliable the target characterization.

Multiple captures can be obtained by moving a single antenna to different locations, by placing multiple stationary antennas at different locations, or combinations thereof.

The advantage of a single moving antenna is that it can be easily placed in any number of positions to provide any number of monostatic waveforms. For example, an antenna mounted on an airplane takes many captures per second as the plane travels.

The principal advantages of multiple static antennas are that a moving target can be characterized (assuming the capture electronics are fast enough), that no vehicle or motion machinery is necessary, and that antenna positions need not be derived from other, sometimes unreliable, information. (One problem with SAR aboard an airplane is knowing precise antenna positions as the plane travels).

For multiple static antennas, all combinations of monostatic and multistatic radar waveform captures are possible. However, that it is not advantageous to capture a waveform for each of both transmission directions for a given pair of antennas, because those waveforms will be identical. When multiple static antennas are used, the total number of unique echo waveforms that can be captured is:

$$\frac{N^2 + N}{2},$$

where N is the number of unique antenna positions.

Modes

Stripmap Mode Airborne SAR

The antenna stays in a fixed position, and may be orthogonal to the flight path or squinted slightly forward or backward.

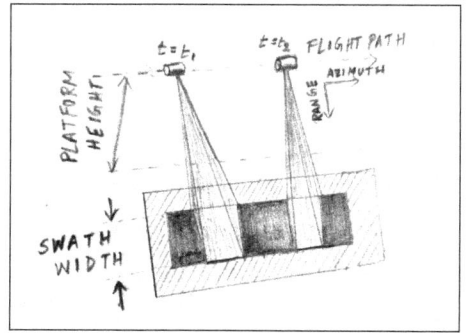

Illustration of the SAR stripmap operation mode.

When the antenna aperture travels along the flight path, a signal is transmitted at a rate equal to the pulse repetition frequency (PRF). The lower boundary of the PRF is determined by the Doppler bandwidth of the radar. The backscatter of each of these signals is commutatively added on a pixel-by-pixel basis to attain the fine azimuth resolution desired in radar imagery.

Spotlight Mode SAR

The spotlight synthetic aperture is given by:

$$Lsa = r_0 \Delta \theta_a$$

where $\Delta \theta_a$ is the angle formed between the beginning and end of the imaging, as shown in the diagram of spotlight imaging and r_0 is the range distance.

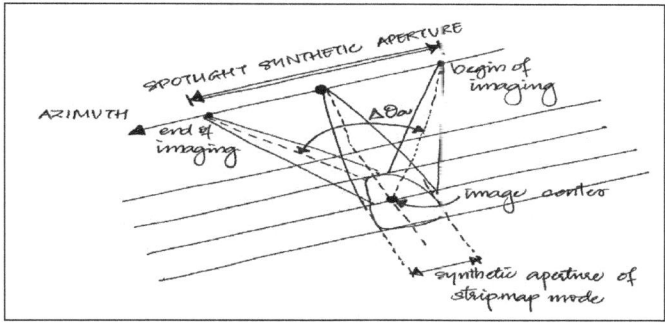

Depiction of the Spotlight Image Mode.

The spotlight mode gives better resolution for a smaller ground patch. In this mode, the illuminating radar beam is steered continually as the aircraft moves, so that it illuminates the same patch over a longer period of time. This mode is not a very continuous imaging mode; however, has high azimuth resolution.

Scan Mode SAR

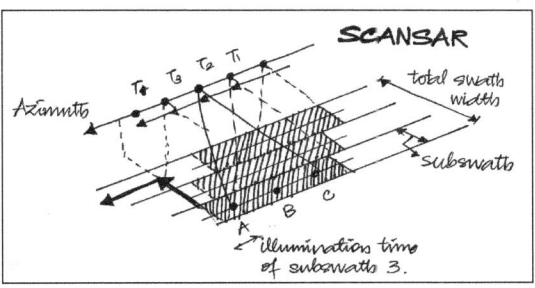

Depiction of ScanSAR Imaging Mode.

While operating as a scan mode SAR, the antenna beam sweeps periodically and thus cover much larger area than the spotlight and stripmap modes. However, the azimuth resolution become much lower than the stripmap mode due to the decreased azimuth bandwidth. Clearly there is a balance achieved between the azimuth resolution and the scan area of SAR. Here, the synthetic aperture is shared between the sub swaths, and it is not in direct contact within one subswath. Mosaic operation is required in azimuth and range directions to join the azimuth bursts and the range sub-swaths.

Properties

- Scan SAR makes the swath beam huge.

- The azimuth signal has many bursts.

- The azimuth resolution is limited due to the burst duration.

- Each target contains varied frequencies which completely depends on where the azimuth is present.

Polarimetry

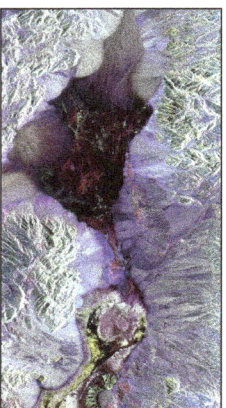

SAR image of Death Valley colored using polarimetry.

Radar waves have a polarization. Different materials reflect radar waves with different intensities, but anisotropic materials such as grass often reflect different polarizations with different intensities. Some materials will also convert one polarization into another. By emitting a mixture of polarizations and using receiving antennas with a specific polarization, several images can be collected from the same series of pulses. Frequently three such RX-TX polarizations (HH-pol, VV-pol, VH-pol) are used as the three color channels in a synthesized image. This is what has been done in the picture at right. Interpretation of the resulting colors requires significant testing of known materials.

New developments in polarimetry include using the changes in the random polarization returns of some surfaces (such as grass or sand) and between two images of the same location at different times to determine where changes not visible to optical systems occurred. Examples include subterranean tunneling or paths of vehicles driving through the area being imaged. Enhanced SAR sea oil slick observation has been developed by appropriate physical modelling and use of fully polarimetric and dual-polarimetric measurements.

SAR polarimetry is a technique used for deriving qualitative and quantitative physical information for land, snow and ice, ocean and urban applications based on the measurement and exploration of the polarimetric properties of man-made and natural scatterers. *Terrain* and *land use* classification is one of the most important applications of polarimetric synthetic-aperture radar (POLSAR).

SAR polarimetry uses a scattering matrixs to identify the scattering behavior of objects after an interaction with electromagnetic wave. The matrix is represented by a combination of horizontal and vertical polarization states of transmitted and received signals.

$$S = \begin{bmatrix} S_{HH} & S_{HV} \\ S_{VH} & S_{VV} \end{bmatrix}$$

where, HH is for horizontal transmit and horizontal receive, VV is for vertical transmit and vertical receive, HV is for horizontal transmit and vertical receive, and VH – for vertical transmit and horizontal receive.

The first two of these polarization combinations are referred to as like-polarized (or co-polarized), because the transmit and receive polarizations are the same. The last two combinations are referred to as cross-polarized because the transmit and receive polarizations are orthogonal to one another.

The three-component scattering power model by Freeman and Durden is successfully used for decomposition of POLSAR image, applying the reflection symmetry condition using covariance matrix. The method is based on simple physical scattering mechanisms (surface scattering, double-bounce scattering, and volume scattering). The advantage of this scattering model is that it is simple and easy to implement for image processing. There are 2 major approaches for a 3×3 polarimetric matrix decomposition. One is the lexicographic covariance matrix approach based on physically measurable parameters, and the other is the Pauli decomposition which is a coherent decomposition matrix. It represents all the polarimetric information in a single SAR image. The polarimetric information of [S] could be represented by the combination of the intensities $|S_{HH}|^2, |S_{VV}|^2, 2|S_{HV}|^2$ in a single RGB image where all the previous intensities will be coded as a color channel.

For PolSAR image analysis, there can be cases where reflection symmetry condition does not hold. In those cases a *four-component scattering model* can be used to decompose polarimetric synthetic-aperture radar (SAR) images. This approach deals with the non- reflection symmetric scattering case. It includes and extends the three-component decomposition method introduced by Freeman and Durden to a fourth component by adding the helix scattering power. This helix power term generally appears in complex urban area but disappears for a natural distributed scatterer.

There is also an improved method using the four-component decomposition algorithm, which was introduced for the general POLSAR data image analyses. The SAR data is first filtered which is known as speckle reduction, then each pixel is decomposed by four-component model to determine the surface scattering power (P_s), double-bounce scattering power (P_s), volume scattering power (P_d), and helix scattering power (P_v).The pixels are then divided into 5 classes (surface,double-bounce,volume,helix,and mixed pixels) classified with respect to maximum powers. A mixed category is added for the pixels having two or three equal dominant scattering powers after computation. The process continues as the pixels in all these categories are divided in 20 small clutter approximately of same number of pixels and merged as desirable, this is called cluster merging. They are iteratively classified and then automatically color is delivered to each class. The summarization of this algorithm leads to an understanding that, brown colors denotes the surface scattering classes, red colors for double-bounce scattering classes, green colors for volume scattering classes, and blue colors for helix scattering classes.

Color representation of different polarizations.

Although this method is aimed for non-reflection case, it automatically includes the reflection symmetry condition, therefore in can be used as a general case. It also preserves the scattering characteristics by taking the mixed scattering category into account therefore proving to be a better algorithm.

Interferometry

Rather than discarding the phase data, information can be extracted from it. If two observations of the same terrain from very similar positions are available, aperture synthesis can be performed to provide the resolution performance which would be given by a radar system with dimensions equal to the separation of the two measurements. This technique is called interferometric SAR or InSAR.

If the two samples are obtained simultaneously (perhaps by placing two antennas on the same aircraft, some distance apart), then any phase difference will contain information about the angle from which the radar echo returned. Combining this with the distance information, one can determine the position in three dimensions of the image pixel. In other words, one can extract terrain altitude as well as radar reflectivity, producing a digital elevation model (DEM) with a single airplane pass. One aircraft application at the Canada Centre for Remote Sensing produced digital elevation maps with a resolution of 5 m and altitude errors also about 5 m. Interferometry was used to map many regions of the Earth's surface with unprecedented accuracy using data from the Shuttle Radar Topography Mission.

If the two samples are separated in time, perhaps from two flights over the same terrain, then there are two possible sources of phase shift. The first is terrain altitude, The second is terrain motion: if the terrain has shifted between observations, it will return a different phase. The amount of shift required to cause a significant phase difference is on the order of the wavelength used. This means that if the terrain shifts by centimeters, it can be seen in the resulting image (a digital elevation map must be available to separate the two kinds of phase difference; a third pass may be necessary to produce one).

This second method offers a powerful tool in geology and geography. Glacier flow can be mapped with two passes. Maps showing the land deformation after a minor earthquake or after a volcanic eruption (showing the shrinkage of the whole volcano by several centimeters) have been published (where?).

Differential Interferometry

Differential interferometry (D-InSAR) requires taking at least two images with addition of a DEM. The DEM can be either produced by GPS measurements or could be generated by interferometry as long as the time between acquisition of the image pairs is short, which guarantees minimal distortion of the image of the target surface. In principle, 3 images of the ground area with similar image acquisition geometry is often adequate for D-InSar. The principle for detecting ground movement is quite simple. One interferogram is created from the first two images; this is also called the reference interferogram or topographical interferogram. A second interferogram is created that captures topography + distortion. Subtracting the latter from the reference interferogram can reveal differential fringes, indicating movement. The described 3 image D-InSAR generation technique is called 3-pass or double-difference method.

Differential fringes which remain as fringes in the differential interferogram are a result of SAR range changes of any displaced point on the ground from one interferogram to the next. In the differential interferogram, each fringe is directly proportional to the SAR wavelength, which is about 5.6 cm for ERS and RADARSAT single phase cycle. Surface displacement away from the satellite look direction causes an increase in path (translating to phase) difference. Since the signal travels from the SAR antenna to the target and back again, the measured displacement is twice the unit of wavelength. This means in differential interferometry one fringe cycle $-\pi$ to $+\pi$ or one wavelength corresponds to a displacement relative to SAR antenna of only half wavelength (2.8 cm). There are various publications on measuring subsidence movement, slope stability analysis, landslide, glacier movement, etc. tooling D-InSAR. Further advancement to this technique whereby differential interferometry from satellite SAR ascending pass and descending pass can be used to estimate 3-D ground movement. Research in this area has shown accurate measurements of 3-D ground movement with accuracies comparable to GPS based measurements can be achieved.

Tomo-SAR

SAR Tomography is a subfield of a concept named as multi-baseline interferometry. It has been developed to give a 3D exposure to the imaging, which uses the beam formation concept. It can be used when the use demands a focused phase concern between the magnitude and the phase components of the SAR data, during information retrieval. One of the major advantages of Tomo-SAR is that it can separate out the parameters which get scattered, irrespective of how different their motions are.

On using Tomo-SAR with differential interferometry, a new combination named "differential tomography" (Diff-Tomo) is developed.

Application of Tomo-SAR

Tomo-SAR has an application based on radar imaging, which is the depiction of Ice Volume and Forest Temporal Coherence (Temporal coherence describes the correlation between waves observed at different moments in time).

Ultra-wideband SAR

Conventional radar systems emit bursts of radio energy with a fairly narrow range of frequencies. A narrow-band channel, by definition, does not allow rapid changes in modulation. Since it is the change in a received signal that reveals the time of arrival of the signal (obviously an unchanging signal would reveal nothing about "when" it reflected from the target), a signal with only a slow change in modulation cannot reveal the distance to the target as well as can a signal with a quick change in modulation.

Ultra-wideband (UWB) refers to any radio transmission that uses a very large bandwidth – which is the same as saying it uses very rapid changes in modulation. Although there is no set bandwidth value that qualifies a signal as "UWB", systems using bandwidths greater than a sizable portion of the center frequency (typically about ten percent, or so) are most often called "UWB" systems. A typical UWB system might use a bandwidth of one-third to one-half of its center frequency. For example, some systems use a bandwidth of about 1 GHz centered around 3 GHz.

There are as many ways to increase the bandwidth of a signal as there are forms of modulation – it is simply a matter of increasing the rate of that modulation. However, the two most common methods used in UWB radar, including SAR, are very short pulses and high-bandwidth chirping. A general description of chirping appears elsewhere in this article. The bandwidth of a chirped system can be as narrow or as wide as the designers desire. Pulse-based UWB systems, being the more common method associated with the term "UWB radar".

A pulse-based radar system transmits very short pulses of electromagnetic energy, typically only a few waves or less. A very short pulse is, of course, a very rapidly changing signal, and thus occupies a very wide bandwidth. This allows far more accurate measurement of distance, and thus resolution.

The main disadvantage of pulse-based UWB SAR is that the transmitting and receiving front-end electronics are difficult to design for high-power applications. Specifically, the transmit duty cycle is so exceptionally low and pulse time so exceptionally short, that the electronics must be capable of extremely high instantaneous power to rival the average power of conventional radars. (Although it is true that UWB provides a notable gain in channel capacity over a narrow band signal because of the relationship of bandwidth in the Shannon–Hartley theorem and because the low receive duty cycle receives less noise, increasing the signal-to-noise ratio, there is still a notable disparity in link budget because conventional radar might be several orders of magnitude more powerful than a typical pulse-based radar.) So pulse-based UWB SAR is typically used in applications requiring average power levels in the microwatt or milliwatt range, and thus is used for scanning smaller, nearer target areas (several tens of meters), or in cases where lengthy integration (over a span of minutes) of the received signal is possible. Note, however, that this limitation is solved in chirped UWB radar systems.

The principal advantages of UWB radar are better resolution (a few millimeters using commercial off-the-shelf electronics) and more spectral information of target reflectivity.

Doppler-beam Sharpening

Doppler Beam Sharpening commonly refers to the method of processing unfocused real-beam phase history to achieve better resolution than could be achieved by processing the real beam without it. Because the real aperture of the radar antenna is so small (compared to the wavelength in use), the radar energy spreads over a wide area (usually many degrees wide in a direction orthogonal (at right angles) to the direction of the platform (aircraft)). Doppler-beam sharpening takes advantage of the motion of the platform in that targets ahead of the platform return a Doppler upshifted signal (slightly higher in frequency) and targets behind the platform return a Doppler downshifted signal (slightly lower in frequency).

The amount of shift varies with the angle forward or backward from the ortho-normal direction. By knowing the speed of the platform, target signal return is placed in a specific angle "bin" that changes over time. Signals are integrated over time and thus the radar "beam" is synthetically reduced to a much smaller aperture – or more accurately (and based on the ability to distinguish smaller Doppler shifts) the system can have hundreds of very "tight" beams concurrently. This technique dramatically improves angular resolution; however, it is far more difficult to take advantage of this technique for range resolution.

Chirped (Pulse-compressed) Radars

A common technique for many radar systems (usually also found in SAR systems) is to "chirp" the signal. In a "chirped" radar, the pulse is allowed to be much longer. A longer pulse allows more energy to be emitted, and hence received, but usually hinders range resolution. But in a chirped radar, this longer pulse also has a frequency shift during the pulse (hence the chirp or frequency shift). When the "chirped" signal is returned, it must be correlated with the sent pulse. Classically, in analog systems, it is passed to a dispersive delay line (often a surface acoustic wave device) that has the property of varying velocity of propagation based on frequency. This technique "compresses" the pulse in time – thus having the effect of a much shorter pulse (improved range resolution) while having the benefit of longer pulse length (much more signal returned). Newer systems use digital pulse correlation to find the pulse return in the signal.

Typical Operation

In a typical SAR application, a single radar antenna is attached to an aircraft or spacecraft so as to radiate a beam whose wave-propagation direction has a substantial component perpendicular to the flight-path direction. The beam is allowed to be broad in the vertical direction so it will illuminate the terrain from nearly beneath the aircraft out toward the horizon.

NASA's Air SAR instrument is attached to the side of a DC-8.

Resolution in the range dimension of the image is accomplished by creating pulses which define very short time intervals, either by emitting short pulses consisting of a carrier frequency and the necessary sidebands, all within a certain bandwidth, or by using longer "chirp pulses" in which frequency varies (often linearly) with time within that bandwidth. The differing times at which echoes return allow points at different distances to be distinguished.

SAR antenna of the SAOCOM satellites.

The total signal is that from a beamwidth-sized patch of the ground. To produce a beam that is narrow in the cross-range direction, diffraction effects require that the antenna be wide in that dimension. Therefore, the distinguishing, from each other, of co-range points simply by strengths of returns that persist for as long as they are within the beam width is difficult with aircraft-carryable antennas, because their beams can have linear widths only about two orders of magnitude (hundreds of times) smaller than the range. (Spacecraft-carryable ones can do 10 or more times better.) However, if both the amplitude and the phase of returns are recorded, then the portion of that multi-target return that was scattered radially from any smaller scene element can be extracted by phase-vector correlation of the total return with the form of the return expected from each such element. Careful design and operation can accomplish resolution of items smaller than a millionth of the range, for example, 30 cm at 300 km, or about one foot at nearly 200 miles (320 km).

The process can be thought of as combining the series of spatially distributed observations as if all had been made simultaneously with an antenna as long as the beamwidth and focused on that particular point. The "synthetic aperture" simulated at maximum system range by this process not only is longer than the real antenna, but, in practical applications, it is much longer than the radar aircraft, and tremendously longer than the radar spacecraft.

Image resolution of SAR in its range coordinate (expressed in image pixels per distance unit) is mainly proportional to the radio bandwidth of whatever type of pulse is used. In the cross-range coordinate, the similar resolution is mainly proportional to the bandwidth of the Doppler shift of the signal returns within the beamwidth. Since Doppler frequency depends on the angle of the scattering point's direction from the broadside direction, the Doppler bandwidth available within the beamwidth is the same at all ranges. Hence the theoretical spatial resolution limits in both image dimensions remain constant with variation of range. However, in practice, both the errors that accumulate with data-collection time and the particular techniques used in post-processing further limit cross-range resolution at long ranges.

The conversion of return delay time to geometric range can be very accurate because of the natural constancy of the speed and direction of propagation of electromagnetic waves. However, for an aircraft flying through the never-uniform and never-quiescent atmosphere, the relating of pulse transmission and reception times to successive geometric positions of the antenna must be accompanied by constant adjusting of the return phases to account for sensed irregularities in the flight path. SAR's in spacecraft avoid that atmosphere problem, but still must make corrections for known antenna movements due to rotations of the spacecraft, even those that are reactions to movements of onboard machinery. Locating a SAR in a manned space vehicle may require that the humans carefully remain motionless relative to the vehicle during data collection periods.

Although some references to SARs have characterized them as "radar telescopes", their actual optical analogy is the microscope, the detail in their images being smaller than the length of the synthetic aperture. In radar-engineering terms, while the target area is in the "far field" of the illuminating antenna, it is in the "near field" of the simulated one.

Returns from scatterers within the range extent of any image are spread over a matching time interval. The inter-pulse period must be long enough to allow farthest-range returns from any pulse to finish arriving before the nearest-range ones from the next pulse begin to appear, so that those do not overlap each other in time. On the other hand, the interpulse rate must be fast enough to provide sufficient samples for the desired across-range (or across-beam) resolution. When the radar is to be carried by a high-speed vehicle and is to image a large area at fine resolution, those conditions may clash, leading to what has been called SAR's ambiguity problem. The same considerations apply to "conventional" radars also, but this problem occurs significantly only when resolution is so fine as to be available only through SAR processes. Since the basis of the problem is the information-carrying capacity of the single signal-input channel provided by one antenna, the only solution is to use additional channels fed by additional antennas. The system then becomes a hybrid of a SAR and a phased array, sometimes being called a Vernier array.

Combining the series of observations requires significant computational resources, usually using Fourier transform techniques. The high digital computing speed now available allows such processing to be done in near-real time on board a SAR aircraft. (There is necessarily a minimum time delay until all parts of the signal have been received.) The result is a map of radar reflectivity, including both amplitude and phase. The amplitude information, when shown in a map-like display, gives information about ground cover in much the same way that a black-and-white photo does. Variations in processing may also be done in either vehicle-borne stations or ground stations for various purposes, so as to accentuate certain image features for detailed target-area analysis.

Although the phase information in an image is generally not made available to a human observer of an image display device, it can be preserved numerically, and sometimes allows certain additional features of targets to be recognized. Unfortunately, the phase differences between adjacent image picture elements ("pixels") also produce random interference effects called "coherence speckle", which is a sort of graininess with dimensions on the order of the resolution, causing the concept of resolution to take on a subtly different meaning. This effect is the same as is apparent both visually and photographically in laser-illuminated optical scenes. The scale of that random speckle structure is governed by the size of the synthetic aperture in wavelengths, and cannot be finer than the system's resolution. Speckle structure can be subdued at the expense of resolution.

Before rapid digital computers were available, the data processing was done using an optical holography technique. The analog radar data were recorded as a holographic interference pattern on photographic film at a scale permitting the film to preserve the signal bandwidths (for example, 1:1,000,000 for a radar using a 0.6-meter wavelength). Then light using, for example, 0.6-micrometer waves (as from a helium–neon laser) passing through the hologram could project a terrain image at a scale recordable on another film at reasonable processor focal distances of around a meter. This worked because both SAR and phased arrays are fundamentally similar to optical holography, but using microwaves instead of light waves. The "optical data-processors" developed for this radar purpose were the first effective analog optical computer systems, and were, in fact, devised before the holographic technique was fully adapted to optical imaging. Because of the different sources of range and across-range signal structures in the radar signals, optical data-processors for SAR included not only both spherical and cylindrical lenses, but sometimes conical ones.

Image Appearance

The following considerations apply also to real-aperture terrain-imaging radars, but are more consequential when resolution in range is matched to a cross-beam resolution that is available only from a SAR.

The two dimensions of a radar image are range and cross-range. Radar images of limited patches of terrain can resemble oblique photographs, but not ones taken from the location of the radar. This is because the range coordinate in a radar image is perpendicular to the vertical-angle coordinate of an oblique photo. The apparent entrance-pupil position (or camera center) for viewing such an image is therefore not as if at the radar, but as if at a point from which the viewer's line of sight is perpendicular to the slant-range direction connecting radar and target, with slant-range increasing from top to bottom of the image.

Because slant ranges to level terrain vary in vertical angle, each elevation of such terrain appears as a curved surface, specifically a hyperbolic cosine one. Verticals at various ranges are perpendiculars to those curves. The viewer's apparent looking directions are parallel to the curve's "hypcos" axis. Items directly beneath the radar appear as if optically viewed horizontally (i.e., from the side) and those at far ranges as if optically viewed from directly above. These curvatures are not evident unless large extents of near-range terrain, including steep slant ranges, are being viewed.

When viewed as specified above, fine-resolution radar images of small areas can appear most nearly like familiar optical ones, for two reasons. The first reason is easily understood by imagining a flagpole in the scene. The slant-range to its upper end is less than that to its base. Therefore, the pole can appear correctly top-end up only when viewed in the above orientation. Secondly, the radar illumination then being downward, shadows are seen in their most-familiar "overhead-lighting" direction.

Note that the image of the pole's top will overlay that of some terrain point which is on the same slant range arc but at a shorter horizontal range ("ground-range"). Images of scene surfaces which faced both the illumination and the apparent eyepoint will have geometries that resemble those of an optical scene viewed from that eyepoint. However, slopes facing the radar will be foreshortened and ones facing away from it will be lengthened from their horizontal (map) dimensions. The former will therefore be brightened and the latter dimmed.

Returns from slopes steeper than perpendicular to slant range will be overlaid on those of lower-elevation terrain at a nearer ground-range, both being visible but intermingled. This is especially the case for vertical surfaces like the walls of buildings. Another viewing inconvenience that arises when a surface is steeper than perpendicular to the slant range is that it is then illuminated on one face but "viewed" from the reverse face. Then one "sees", for example, the radar-facing wall of a building as if from the inside, while the building's interior and the rear wall (that nearest to, hence expected to be optically visible to, the viewer) have vanished, since they lack illumination, being in the shadow of the front wall and the roof. Some return from the roof may overlay that from the front wall, and both of those may overlay return from terrain in front of the building. The visible building shadow will include those of all illuminated items. Long shadows may exhibit blurred edges due to the illuminating antenna's movement during the "time exposure" needed to create the image.

Surfaces that we usually consider rough will, if that roughness consists of relief less than the radar wavelength, behave as smooth mirrors, showing, beyond such a surface, additional images of items in front of it. Those mirror images will appear within the shadow of the mirroring surface, sometimes filling the entire shadow, thus preventing recognition of the shadow.

An important fact that applies to SARs but not to real-aperture radars is that the direction of overlay of any scene point is not directly toward the radar, but toward that point of the SAR's current path direction that is nearest to the target point. If the SAR is "squinting" forward or aft away from the exactly broadside direction, then the illumination direction, and hence the shadow direction, will not be opposite to the overlay direction, but slanted to right or left from it. An image will appear with the correct projection geometry when viewed so that the overlay direction is vertical, the SAR's flight-path is above the image, and range increases somewhat downward.

Objects in motion within a SAR scene alter the Doppler frequencies of the returns. Such objects therefore appear in the image at locations offset in the across-range direction by amounts proportional to the range-direction component of their velocity. Road vehicles may be depicted off the roadway and therefore not recognized as road traffic items. Trains appearing away from their tracks are more easily properly recognized by their length parallel to known trackage as well as by the absence of an equal length of railbed signature and of some adjacent terrain, both having been shadowed by the train. While images of moving vessels can be offset from the line of the earlier parts of their wakes, the more recent parts of the wake, which still partake of some of the vessel's motion, appear as curves connecting the vessel image to the relatively quiescent far-aft wake. In such identifiable cases, speed and direction of the moving items can be determined from the amounts of their offsets. The along-track component of a target's motion causes some defocus. Random motions such as that of wind-driven tree foliage, vehicles driven over rough terrain, or humans or other animals walking or running generally render those items not focusable, resulting in blurring or even effective invisibility.

These considerations, along with the speckle structure due to coherence, take some getting used to in order to correctly interpret SAR images. To assist in that, large collections of significant target signatures have been accumulated by performing many test flights over known terrains and cultural objects.

Relationship to Phased Arrays

A technique closely related to SAR uses an array (referred to as a "phased array") of real antenna elements spatially distributed over either one or two dimensions perpendicular to the radar-range dimension. These physical arrays are truly synthetic ones, indeed being created by synthesis of a collection of subsidiary physical antennas. Their operation need not involve motion relative to targets. All elements of these arrays receive simultaneously in real time, and the signals passing through them can be individually subjected to controlled shifts of the phases of those signals. One result can be to respond most strongly to radiation received from a specific small scene area, focusing on that area to determine its contribution to the total signal received. The coherently detected set of signals received over the entire array aperture can be replicated in several data-processing channels and processed differently in each. The set of responses thus traced to different small scene areas can be displayed together as an image of the scene.

In comparison, a SAR's (commonly) single physical antenna element gathers signals at different positions at different times. When the radar is carried by an aircraft or an orbiting vehicle, those positions are functions of a single variable, distance along the vehicle's path, which is a single mathematical dimension (not necessarily the same as a linear geometric dimension). The signals are stored, thus becoming functions, no longer of time, but of recording locations along that dimension. When the stored signals are read out later and combined with specific phase shifts, the result is the same as if the recorded data had been gathered by an equally long and shaped phased array. What is thus synthesized is a set of signals equivalent to what could have been received simultaneously by such an actual large-aperture (in one dimension) phased array. The SAR simulates (rather than synthesizes) that long one-dimensional phased array.

While operation of a phased array is readily understood as a completely geometric technique, the fact that a synthetic aperture system gathers its data as it (or its target) moves at some speed means that phases which varied with the distance traveled originally varied with time, hence constituted temporal frequencies. Temporal frequencies being the variables commonly used by radar engineers, their analyses of SAR systems are usually (and very productively) couched in such terms. In particular, the variation of phase during flight over the length of the synthetic aperture is seen as a sequence of Doppler shifts of the received frequency from that of the transmitted frequency. It is significant, though, to realize that, once the received data have been recorded and thus have become timeless, the SAR data-processing situation is also understandable as a special type of phased array, treatable as a completely geometric process.

The core of both the SAR and the phased array techniques is that the distances that radar waves travel to and back from each scene element consist of some integer number of wavelengths plus some fraction of a "final" wavelength. Those fractions cause differences between the phases of the re-radiation received at various SAR or array positions. Coherent detection is needed to capture the signal phase information in addition to the signal amplitude information. That type of detection requires finding the differences between the phases of the received signals and the simultaneous phase of a well-preserved sample of the transmitted illumination.

Every wave scattered from any point in the scene has a circular curvature about that point as a center. Signals from scene points at different ranges therefore arrive at a planar array with different curvatures, resulting in signal phase changes which follow different quadratic variations across a planar phased array. Additional linear variations result from points located in different directions from the center of the array. Fortunately, any one combination of these variations is unique to one scene point, and is calculable. For a SAR, the two-way travel doubles that phase change.

In reading the following two paragraphs, be particularly careful to distinguish between array elements and scene elements. Also remember that each of the latter has, of course, a matching image element.

Comparison of the array-signal phase variation across the array with the total calculated phase variation pattern can reveal the relative portion of the total received signal that came from the only scene point that could be responsible for that pattern. One way to do the comparison is by a correlation computation, multiplying, for each scene element, the received and the calculated field-intensity values array element by array element and then summing the products for each scene element. Alternatively, one could, for each scene element, subtract each array element's calculated phase shift from the actual received phase and then vectorially sum the resulting field-intensity differences over the array. Wherever in the scene the two phases substantially cancel everywhere in the array, the difference vectors being added are in phase, yielding, for that scene point, a maximum value for the sum.

The equivalence of these two methods can be seen by recognizing that multiplication of sinusoids can be done by summing phases which are complex-number exponents of e, the base of natural logarithms.

However it is done, the image-deriving process amounts to "backtracking" the process by which nature previously spread the scene information over the array. In each direction, the process may be viewed as a Fourier transform, which is a type of correlation process. The image-extraction process we use can then be seen as another Fourier transform which is a reversal of the original natural one.

It is important to realize that only those sub-wavelength differences of successive ranges from the transmitting antenna to each target point and back, which govern signal phase, are used to refine the resolution in any geometric dimension. The central direction and the angular width of the illuminating beam do not contribute directly to creating that fine resolution. Instead, they serve only to select the solid-angle region from which usable range data are received. While some distinguishing of the ranges of different scene items can be made from the forms of their sub-wavelength range variations at short ranges, the very large depth of focus that occurs at long ranges usually requires that over-all range differences (larger than a wavelength) be used to define range resolutions comparable to the achievable cross-range resolution.

Data Collection

Highly accurate data can be collected by aircraft overflying the terrain in question. In the 1980s, as a prototype for instruments to be flown on the NASA Space Shuttles, NASA operated a synthetic aperture radar on a NASA Convair 990. In 1986, this plane caught fire on takeoff. In 1988, NASA rebuilt a C, L, and P-band SAR to fly on the NASA DC-8 aircraft. Called AIRSAR, it flew missions at sites around the world until 2004. Another such aircraft, the Convair 580, was flown by the Canada Center for Remote Sensing until about 1996 when it was handed over to Environment Canada due to budgetary reasons. Most land-surveying applications are now carried out by satellite observation. Satellites such as ERS-1/2, JERS-1, Envisat ASAR, and RADARSAT-1 were launched explicitly to carry out this sort of observation. Their capabilities differ, particularly in their support for interferometry, but all have collected tremendous amounts of valuable data. The Space Shuttle also carried synthetic aperture radar equipment during the SIR-A and SIR-B missions during the 1980s, the Shuttle Radar Laboratory (SRL) missions in 1994 and the Shuttle Radar Topography Mission in 2000.

A model of a German SAR-Lupe reconnaissance
satellite inside a Cosmos-3M rocket.

The Venera 15 and Venera 16 followed later by the Magellan space probe mapped the surface of Venus over several years using synthetic aperture radar.

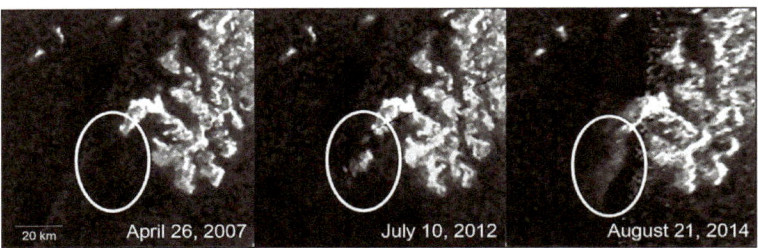

Titan – Evolving feature in Ligeia Mare.

Synthetic aperture radar was first used by NASA on JPL's Seasat oceanographic satellite in 1978 (this mission also carried an altimeter and a scatterometer); it was later developed more extensively on the Spaceborne Imaging Radar (SIR) missions on the space shuttle in 1981, 1984 and 1994. The Cassini mission to Saturn used SAR to map the surface of the planet's major moon Titan, whose surface is partly hidden from direct optical inspection by atmospheric haze. The SHARAD sounding radar on the Mars Reconnaissance Orbiter and MARSIS instrument on Mars Express have observed bedrock beneath the surface of the Mars polar ice and also indicated the likelihood of substantial water ice in the Martian middle latitudes. The Lunar Reconnaissance Orbiter, launched in 2009, carries a SAR instrument called Mini-RF, which was designed largely to look for water ice deposits on the poles of the Moon.

The Mineseeker Project is designing a system for determining whether regions contain landmines based on a blimp carrying ultra-wideband synthetic aperture radar. Initial trials show promise; the radar is able to detect even buried plastic mines.

SAR has been used in radio astronomy for many years to simulate a large radio telescope by combining observations taken from multiple locations using a mobile antenna.

The National Reconnaissance Office maintains a fleet of (now declassified) synthetic aperture radar satellites commonly designated as Lacrosse or Onyx.

Titan – Ligeia Mare – SAR and clearer despeckled views.

In February 2009, the Sentinel R1 surveillance aircraft entered service in the RAF, equipped with the SAR-based Airborne Stand-Off Radar (ASTOR) system.

The German Armed Forces' (Bundeswehr) military SAR-Lupe reconnaissance satellite system has been fully operational since 22 July 2008.

Interferometric Synthetic-aperture Radar

Interferometric Synthetic Aperture Radar (InSAR) is a geodetic technique that can identify movements of the Earth's surface. Observations of surface movement made using InSAR can be used to detect, measure, and monitor crustal changes associated with geophysical processes such as tectonic activity and volcanic eruptions. Ground subsidence caused by anthropogenic influences such as groundwater or hydrocarbon extraction can also be identified with InSAR. When combined with ground-based geodetic monitoring, such as Global Navigation Satellite Systems, InSAR can identify surface movements of millimetre to centimetre scale with high spatial resolution.

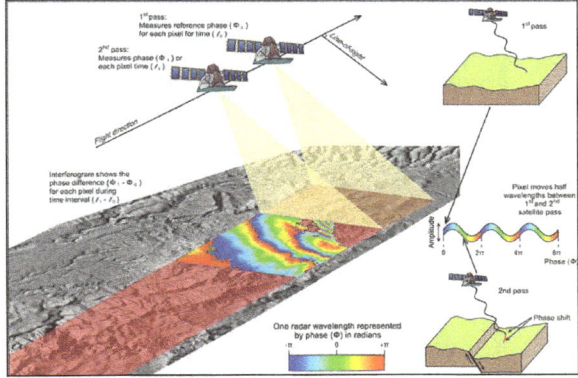

Two SAR images of the same area are acquired at different times. If the surface moves between the two acquisitions a phase shift is recorded. An interferogram maps this phase shift spatially.

InSAR can be used for a wide range of surface deformation studies, for example:

- Subsidence and uplift induced by anthropogenic activities such as groundwater or hydrocarbon extraction, or reinjection into reservoirs during carbon capture and storage.

- Coseismic deformation caused during an earthquake.

- Postseismic and interseismic deformation on crustal faults between earthquakes.

- Inflation/Deflation of subsurface magma chambers preceding volcanic eruptions.

- Monitoring surface movements in urban environments.

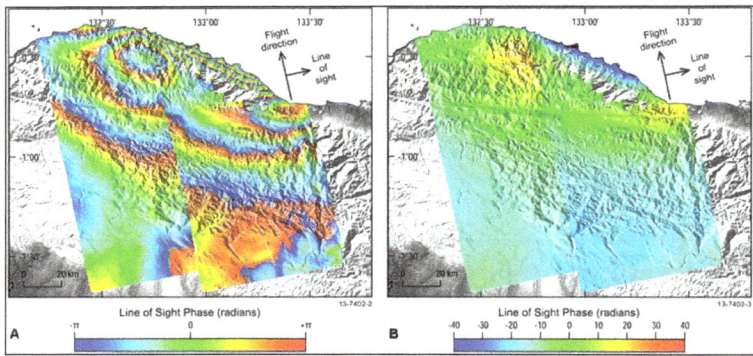

A wrapped (A) and unwrapped (B) interferogram of an earthquake doublet that occurred in West Papua, Indonesia created using data from the Japanese ALOS satellite. The magnitude 7.6 and 7.4 earthquakes occurred on 03 January 2009 within 3 hours of each other and were caused by subduction on the offshore Manokwari Trench, which is located north of the coastline. Unwrapped phase in radians can be converted to 'range change' or displacement in millimetres with knowledge of the satellite radar wavelength.

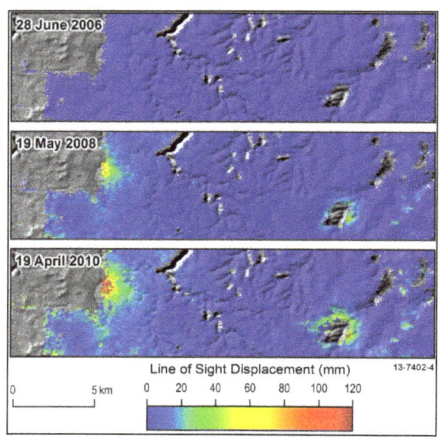

InSAR time-series product showing cumulative surface displacement over time for a small region in the southern New South Wales coal fields. The 1-dimensional displacement observations are in the line-of-sight of the satellite; the slanted path between the ground and satellite position. The positive polarity of the signal at two anomalous zones indicates movement away from the satellite.

How InSAR Works?

InSAR uses two or more Synthetic Aperture Radar (SAR) images of an area to identify surface movements through time. Remote sensing satellites that collect SAR imagery transmit pulses of microwave energy to the Earth's surface and record the amount of backscattered energy. The use of microwave energy provides an all-weather capability because of its low sensitivity to clouds and rain.

SAR images contain information on the Earth's surface in the form of the amplitude and phase components of the backscattered radar signal. The amplitude image records information on the terrain slope and surface roughness, while the phase image records information on the distance between the satellite and the Earth's surface.

Differential InSAR uses two SAR images of the same area acquired at different times. If the distance between the ground and satellite changes between the two acquisitions due to surface movement, a phase shift will occur.

When mapped spatially the phase shift is a 'wrapped' signal within a range of $2\dot\iota$ radians that appears as a series of interference fringes in an interferogram. When this interferogram is unwrapped, the number of fringes is summed to give a continuous field of relative phase change. When first processed, the initial interferogram contains a number of signal components, such as residual signals due to the orbital geometry of the satellite and signals due to the different atmospheric conditions at the time of the two acquisitions. After processing a network of interferograms, the signal component originating from surface movement can be isolated.

InSAR Products

By combining a network of multiple interferograms over a region, velocity map and time-series products can be generated. A velocity map gives the surface movement for each image pixel averaged over the total observation period whereas the time-series product shows the history of surface positions for a pixel at each acquisition time. The former is useful for mapping geophysical processes that occur steadily through time, for example the build-up of strain at a locked crustal fault zone. The latter is useful for detecting geophysical processes that vary considerably through time and may cause fluctuations in the direction of surface movement, for example the inflation and deflation of a magma chamber beneath an active volcano.

Imaging Radar

A SAR radar image acquired by the SIR-C/X-SAR radar on board the Space Shuttle Endeavour shows the Teide volcano. The city of Santa Cruz de Tenerife is visible as the purple and white area on the lower right edge of the island. Lava flows at the summit crater appear in shades of green and brown, while vegetation zones appear as areas of purple, green and yellow on the volcano's flanks.

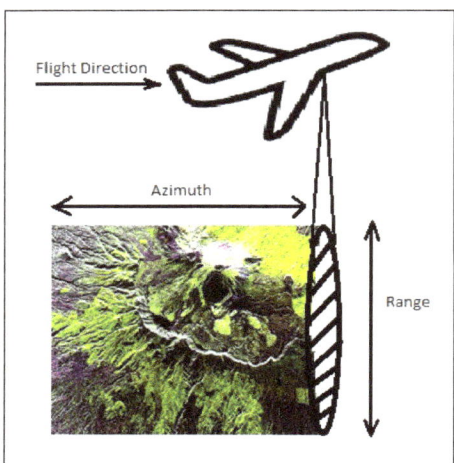

Building up a radar image using the motion of the platform.

Imaging radar is an application of radar which is used to create two-dimensional images, typically of landscapes. Imaging radar provides its light to illuminate an area on the ground and take a picture at radio wavelengths. It uses an antenna and digital computer storage to record its images. In a radar image, one can see only the energy that was reflected back towards the radar antenna. The radar moves along a flight path and the area illuminated by the radar, or footprint, is moved along the surface in a swath, building the image as it does so.

Digital radar images are composed of many dots. Each pixel in the radar image represents the radar backscatter for that area on the ground: brighter areas represent high backscatter, darker areas represents low backscatter.

The traditional application of radar is to display the position and motion of typically highly reflective objects (such as aircraft or ships) by sending out a radiowave signal, and then detecting the direction and delay of the reflected signal. Imaging radar on the other hand attempts to form an image of one object (e.g. a landscape) by furthermore registering the intensity of the reflected signal to determine the amount of scattering (cf. Light scattering). The registered electromagnetic scattering is then mapped onto a two-dimensional plane, with points with a higher reflectivity getting assigned usually a brighter color, thus creating an image.

Several techniques have evolved to do this. Generally they take advantage of the Doppler effect caused by the rotation or other motion of the object and by the changing view of the object brought about by the relative motion between the object and the back-scatter that is perceived by the radar of the object (typically, a plane) flying over the earth. Through recent improvements of the techniques, radar imaging is getting more accurate. Imaging radar has been used to map the Earth, other planets, asteroids, other celestial objects and to categorize targets for military systems.

An imaging radar is a kind of radar equipment which can be used for imaging. A typical radar technology includes emitting radio waves, receiving their reflection, and using this information to generate data. For an imaging radar, the returning waves are used to create an image. When the radio waves reflect off objects, this will make some changes in the radio waves and can provide data about the objects, including how far the waves traveled and what kind of objects they encountered. Using the acquired data, a computer can create a 3-D or 2-D image of the target.

Imaging radar has several advantages. It can operate in the presence of obstacles that obscure the target, and can penetrate ground (sand), water, or walls.

Applications

Applications include: Surface topography & costal change; land use monitoring, agricultural monitoring, ice patrol, environmental monitoring;weather radar- storm monitoring, wind shear warning;medical microwave tomography; through wall radar imaging; 3-D measurements, etc.

Through Wall Radar Imaging

Wall parameter estimation uses Utra Wide-Band radar systems. The handle M-sequence UWB radar with horn and circular antennas was used for data gathering and supporting the scanning method.

3-D Measurements

3-D measurements are supplied by amplitude-modulated laser radars—Erim sensor and Perceptron sensor. In terms of speed and reliability for median-range operations, 3-D measurements have superior performance.

Techniques and Methods

Current radar imaging techniques rely mainly on synthetic aperture radar (SAR) and inverse synthetic aperture radar (ISAR) imaging. Emerging technology utilizes mono-pulse radar 3-D imaging.

Real Aperture Radar

Real aperture radar (RAR) is a form of radar that transmits a narrow angle beam of pulse radio wave in the range direction at right angles to the flight direction and receives the backscattering from the targets which will be transformed to a radar image from the received signals.

Usually the reflected pulse will be arranged in the order of return time from the targets, which corresponds to the range direction scanning.

The resolution in the range direction depends on the pulse width. The resolution in the azimuth direction is identical to the multiplication of beam width and the distance to a target.

AVTIS Radar

The AVTIS radar is a 94 GHz real aperture 3D imaging radar. It uses Frequency-Modulated Continuous-Wave(FMCW) modulation and employs a mechanically scanned monostatic with sub-metre range resolution.

Laser Radar

Laser radar is a remote sensing technology that measures distance by illuminating a target with a laser and analyzing the reflected light.

Laser radar is used for multi-dimensional imaging and information gathering. In all information gathering modes, lasers that transmit in the eye-safe region are required as well as sensitive receivers at these wavelengths.

3-D imaging requires the capacity to measure the range to the first scatter within every pixel. Hence, an array of range counters is needed. A monolithic approach to an array of range counters is being developed. This technology must be coupled with highly sensitive detectors of eye-safe wavelengths.

To measure Doppler information requires a different type of detection scheme than is used for spatial imaging. The returned laser energy must be mixed with a local oscillator in a heterodyne system to allow extraction of the Doppler shift.

3-D, Multi-wave, Multi-band and Imaging Radar

3-D, Multi-wave and Multi-band, Imaging radar works in one of the two modes - as an arbitrary frequency(kHz-MHz), arbitrary wave radar, and as a C-band analog and digital mode radar. The system architecture of 3-D, Multi-wave and Multi-band, Imaging radar is shown in the figure.

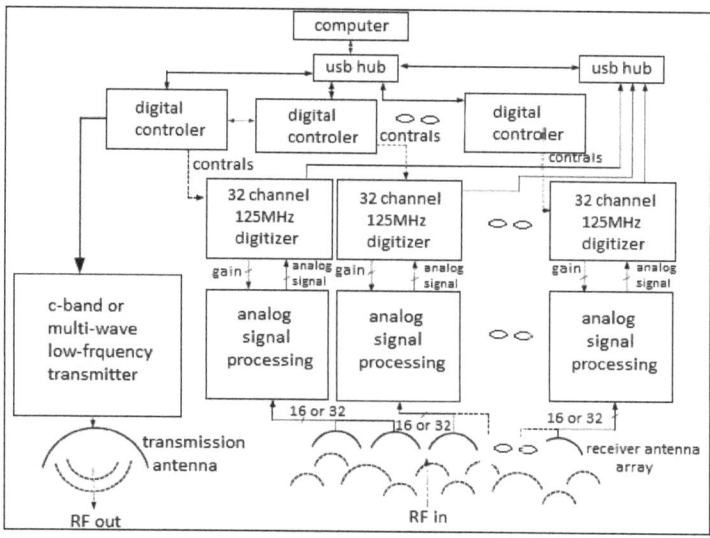

3-D, Multi-wave, Multi-band,Imaging Radar Architecture.

Synthetic Aperture Radar (SAR)

Synthetic aperture radar (SAR) is a form of radar which moves a real aperture or antenna through a series of positions along the objects to provide distinctive long-term coherent-signal variations. This can be used to obtain higher resolution.

SARs produce a two-dimensional (2-D) image. One dimension in the image is called range and is a measure of the "line-of-sight" distance from the radar to the object. Range is determined by measuring the time from transmission of a pulse to receiving the echo from a target. Also, range resolution is determined by the transmitted pulse width.The other dimension is called azimuth and is perpendicular to range. The ability of SAR of producing relatively fine azimuth resolution makes it different from other radars. To obtain fine azimuth resolution, a physically large antenna is needed to focus the transmitted and received energy into a sharp beam. The sharpness of the beam defines the azimuth resolution. An airborne radar could collect data while flying this distance and process the data as if it came from a physically long antenna. The distance the aircraft flies in synthesizing the antenna is known as the synthetic aperture. A narrow synthetic beamwidth results from the relatively long synthetic aperture, which gets finer resolution than a smaller physical antenna.

Inverse Synthetic Aperture Radar (ISAR)

Inverse synthetic aperture radar (ISAR) is another kind of SAR system which can produce high-resolution on two- and three-dimensional images.

An ISAR system consists of a stationary radar antenna and a target scene that is undergoing some motion. ISAR is theoretically equivalent to SAR in that high-azimuth resolution is achieved via relative motion between the sensor and object, yet the ISAR moving target scene is usually made up of non cooperative objects.

Algorithms with more complex schemes for motion error correction are needed for ISAR imaging than those needed in SAR. ISAR technology uses the movement of the target rather than the emitter to make the synthetic aperture. ISAR radars are commonly used on vessels or aircraft and can provide a radar image of sufficient quality for target recognition. The ISAR image is often adequate to discriminate between various missiles, military aircraft, and civilian aircraft.

Disadvantages of ISAR

- The ISAR imaging cannot obtain the real azimuth of the target.

- There sometimes exists a reverse image. For example, the image formed of a boat when it rolls forwards and backwards in the ocean.

- The ISAR image is the 2-D projection image of the target on the Range-Doppler plane which is perpendicular to the rotating axis. When the Range-Doppler plane and the coordinate plane are different, the ISAR image can not reflect the real shape of the target. Thus, the ISAR imaging can not obtain the real shape information of the target in most situations.

Rolling is side to side. Pitching is forward and backwards, yawing is turning left or right.

Monopulse Radar 3-D Imaging Technique

Monopulse radar 3-D imaging technique uses 1-D range image and monopulse angle measurement to get the real coordinates of each scatterer. Using this technique, the image doesn't vary with the change of the target's movement. Monopulse radar 3-D imaging utilizes the ISAR techniques to separate scatterers in the Doppler domain and perform monopulse angle measurement.

Monopulse radar 3-D imaging can obtain the 3 views of 3-D objects by using any two of the three parameters obtained from the azimuth difference beam, elevation difference beam and range measurement, which means the views of front, top and side can be azimuth-elevation, azimuth-range and elevation-range, respectively.

Monopulse imaging generally adapts to near-range targets, and the image obtained by monopulse radar 3-D imaging is the physical image which is consistent with the real size of the object.

A rising tide of acceptance and usage of satellite-derived synthetic aperture radar (SAR) data has occurred during the last few years. This trend is the result of the increasing availability of commercial SAR satellite data; development of sophisticated processing and analysis tools; and industry-driven training initiatives to familiarize image analysts with SAR imagery, including its interpretation and utility.

Electro-optical/SAR Comparisons

Intuitively the colored imagery derived from electro-optical systems provides the human eye with familiar representations of Earth's surface that are instinctively easy to interpret. Additionally, electro-optical imagery has been known to the user community since the 1970s, so there's a lot of know how available.

Regardless, the user community is recognizing there's much more than meets the eye in black-and-white SAR data and imagery. The most obvious SAR advantage is the weather and daylight independence of radar systems, which ensure a guaranteed acquisition of the area of interest. This also enables consistent monitoring independent of lighting, weather or cloud-cover conditions.

This, however, is just one side of the coin. The real advantages of SAR unfold when the data are processed and analyzed appropriately to meet the mission. Many unique effects of SAR satellite data, such as radar shadow, can be exploited to extract information from the derived imagery that isn't detectable through visual interpretation alone.

For example, SAR imagery can be used to detect and even quantify the motion of objects on both land and sea. Thanks to the measurability of a SAR signal's intensity and phase, imagery analysts can determine elevation information and even subtle changes to surface conditions.

Electro-optical systems are passive, which means they require the illumination of the sun for imaging. Radar, however, is an active remote sensing system, which means it provides its own energy source to illuminate the imaging area. A radar imaging system has three main functions: It transmits the microwave signal toward the scene, receives a portion of that transmitted energy as backscatter from the scene, and then observes the strength and time delay of the returned signal.

The energy of the radar pulse is scattered in all directions at the Earth's surface, with some reflected back to the antenna. The surface's roughness—i.e., the irregularity of the terrain vertically and horizontally—determines the return signal's amplitude. Surfaces can be classified as smooth, slightly rough, moderately rough or very rough.

Generally, bright areas in a SAR image are strong reflectors, such as buildings in urban areas, while dark parts of the image represent surfaces that reflect little or no energy, such as water surfaces or oil film on an ocean. Depending on the wavelength of the radar signal, SAR can penetrate forest canopy and Earth surfaces, detecting dielectric features such as metal objects, water, freeze/thaw, salt content, iron oxides and clay in soils. To fully exploit the advantages of SAR data and imagery, key mission-specific collection characteristics and parameters must be understood. The following sections clarify these parameters and characteristics.

Wavelength

As detailed in the table below, Radar remote sensing uses the microwave portion of the electromagnetic spectrum, from a frequency of 0.3 GHz to 300 GHz. Most radar satellites operate at wavelengths between 0.5 cm and 75 cm.

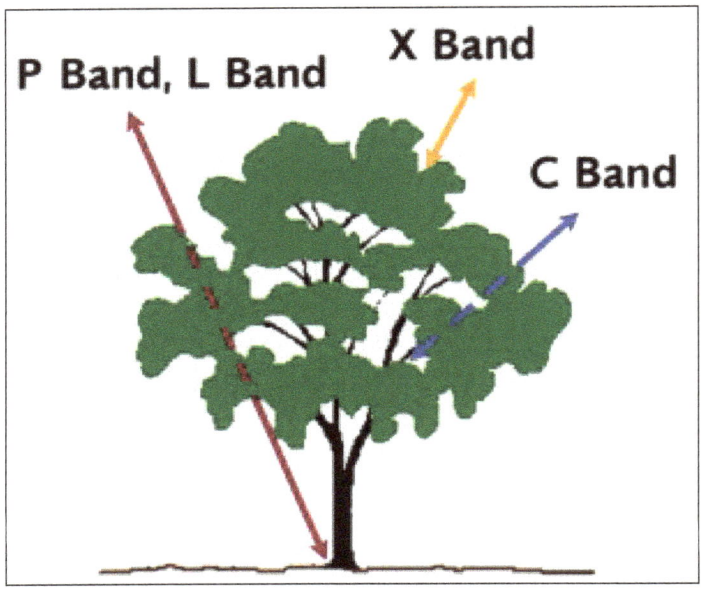

Table: Basic imaging Modes of Commercial Radar Remote Sensing Systems.

		SpotLight	StripMap	ScanSAR
TerraSAR-X	Resolution:	1 meter	3 meter	18 meter
	Scene size:	10 km x 5 km	30 km x 50 metres* *extendable up to 1,650 km	100 km x 150 km* * extendable up to 1,650 km
COS-MO-SkyMed	Resolution:	1 meter	3 metres (HImage) 15 metres (ping pong)	30 metres (Scan SAR Wide) 100 metres (Scan SAR Huge)
	Scene size:	10 km x 10 km	40 km x 40 km (HImage) 30 km x 30 km (ping pong)	100 km x 100 km (ScanSAR Wide) 100 km x 100 km (ScanSAR Huge)

RADAR-SAT-2	Resolution:	~1.6 metres (Spot-Light)	8 metres (Wide Multi-Look Fine)	50 metres (Scan SAR Narrow)
		3 metres (ultra wide fine)	8 metres (Wide Fine)	100 metres (Scan SAR Wide)
	Scene size:	18 km x 8 km (Spot-Light)	90 km x 50 km (Wide Multi-Look Fine)	300 km x 300 Km (Scan SAR Narrow)
		50 km x 50km (ultra wide fine)	150 km x 170 km (Wide Fine)	500 km x 500 km (Scan SAR Wide)

Shorter wavelengths—e.g., X-band imagery at 3 cm—are reflected from the top of the canopy, while longer wavelengths—e.g., L-band imagery at 24 cm—normally go down to the ground and are reflected from there. Using this characteristic of different wavelengths makes it possible to discern information about the canopy structure of a forested area from a multiwavelength image and thus estimate above-ground biomass.

Furthermore, the choice of wavelength needs to be matched to the size of the surface feature that should be distinguishable. Small features are best recognized with X-band imagery—i.e., short wavelengths—while large features, such as geology, are better marked in L-band imagery.

Polarization

Transmitted and received radar signals propagate in a certain plane—the polarization. The propagation planes are usually horizontal (H) and vertical (V). Vertically polarized waves will interact with the vertical stalks of plant canopy, while horizontally polarized waves will penetrate through plant canopy. Thus, the combination of the image channels into a red-green-blue (RGB) image results in a false-color image, which can differentiate ground cover such as vegetation classes.

The different combination options for the polarization will provide different image characteristics:

- Single polarization: The radar system operates with the same polarization for transmitting and receiving the signal.

- Cross polarization: A different polarization is used to transmit and receive the signal.

- Dual polarization: The radar system operates with one polarization to transmit the signal and both polarizations simultaneously to receive the signal.

- Quad polarization: H and V polarizations are used for alternate pulses to transmit the signal and with both simultaneously to receive the signal.

Multipolarized images are provided in the form of multiple layers, each corresponding to a different polarization channel. Each polarization channel is identified by two letters. The first letter denotes the transmit polarization, and the second refers to the receive polarization. Multipolarized SAR imagery allows users to measure the terrain's polarization properties and not simply the backscatter at a single polarization, thus providing improved classification information.

Mode

As detailed in the table below, the acquisition mode is directly linked with the resolution of the resulting image and the size of the scene area covered.

Table: Frequency and Wavelength of Commonly Used Radar Remote Sensing Bands.

Band	Frequency	wavelength	Key Characteristics
X Band	12.5-8 GHz	2.4-3.75 cm	Widely used for military reconnaissance, mapping and surveillance (TerraSAR-X, TanDEM-X, COSMO-SkyMed)
C Band	8-4 GHz	3.75-7.5 cm	Penetration capacity of vegetation or solids is limited and restricted to the top layes, Useful for sea-ice surveillance (RADARSAT, ERS-1)
S Band	4-2 GHz	7.5-15 cm	Used for medium-range meteorological application- eg., rainfall measurement, airport surveillance
L Band	2-1 GHz	15-30 cm	Penetrates vegetation to support observation applications over vegetated surfaces and for monitoring ice sheet and glacier dynamics (ALOS PALSAR)
P Band	1-0.3 GHz	30-100cm	To date only used for research and experimental applications. Significant openetration capabilities regarding vegetation canopy (key element for estimating vegetation biomass), sea ice, soil, glaciers.

The highest resolution is achieved with a SpotLight image. For this the radar beam continuously illuminates one terrain patch while the satellite is moving along its flight path. This sophisticated imaging mode makes it possible to acquire data with up to 1-meter resolution but restricts the scene size.

An image in StripMap mode is acquired by illuminating the ground swath with a continuous sequence of pulses while the antenna beam is pointed to a fixed angle in elevation and azimuth. This results in an image strip with constant image quality in azimuth. StripMap is the most commonly used acquisition mode, as it provides a good trade off between the size of the area covered and the resolution.

The ScanSAR mode overcomes the constraints of the narrow swath of the StripMap and is intended for use in applications requiring large area coverage such as monitoring applications. In this mode, electronic antenna elevation steering is used to acquire adjacent, slightly overlapping coverages with different incidence angles that are processed into one scene. For example, in the case of TerraSAR-X and RADARSAT-2, up to four single beams covering adjoining swaths are combined. Due to the switching between the beams, only bursts of SAR echoes are received, resulting in a reduced bandwidth and hence reduced azimuth resolution.

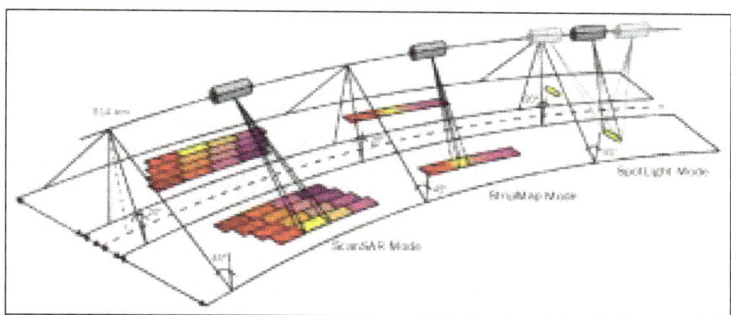

Acquisition mode is directly linked with the resolution of the resulting image and the size of the scene area covered.

Incidence Angle

The incidence angle refers to the angle between the "straight to ground" and the radar illumination. The interaction of microwaves with the ground is complex, and different reflections occur in different angular regions. Returns are normally strong at low incidence angles and decrease with increasing incidence angle.

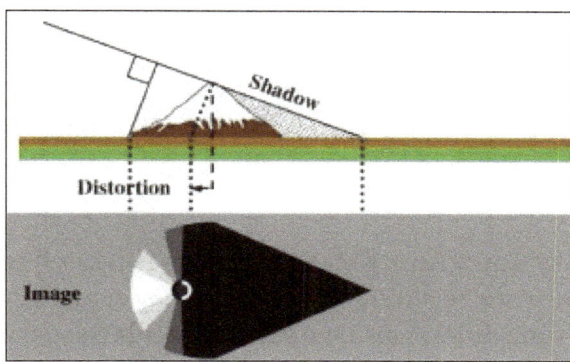

Because of SAR's side-looking perspective, tall objects and relief structures are subject to displacements. There are three main radar effects, which must be taken into account when using SAR data:

- Radar shadows are areas on the ground that aren't illuminated by the radar signal, thus no return signal is received, and these areas appear dark in the imagery. As the incidence angle of an image increases from near-range to far-range, shadowing becomes more prominent toward far-range. Shadowing in a radar image is an important key for terrain relief interpretation, as the height of an object can be derived from measuring the radar shadow. Thus, this apparently negative radar effect provides valuable information about a scene.

- Foreshortening describes the compression appearance of features that are tilt-ed toward the radar. For a given slope, foreshortening effects are reduced with increasing incident angles. With this reduction of incidence angle, however, the shadow effect increases. Thus, the selection of the incidence angle is always a trade-off between the acquisition's occurrence of foreshortening and radar shadow.

- Layover occurs when the reflected signal from a feature's upper portion is re-ceived before the return from the feature's lower portion. In this case, the top of the feature will be displaced relative to its base. This effect is more prevalent for viewing geometries with smaller incident angles.

These effects can be compensated for only through a trade-off among them—if you don't have one, you will have the others. They can, however, support image analysis.

Repeat Frequency

Applications such as interferometry, surface movement monitoring or change detec-tion require the acquisition of data stacks that have identical acquisition parameters (orbit, incidence angle and polarization). Analyzing changes in the amplitude and the phase of the return signal at the various acquisition dates provides information to de-rive small changes on the imaged surface. Not all changes, such as vegetation changes, are relevant for analysis, but the time difference among the acquisitions is important for interferometric processing.

Current commercial radar systems have 24- (RADARSAT-2), 16- (COSMO-SkyMed constellation of four satellites) and 11-day (TerraSAR-X and TanDEM-X) repeat cy-cles—i.e., they pass over the same point on the ground with the identical acquisition geometry in these time intervals.

In the case of TerraSAR-X and TanDEM-X, where the two satellites fly in a close forma-tion with only a distance of a few hundred meters, they can acquire an interferometric data pair without any time difference. One satellite sends the signal, and both satellites record the backscatters simultaneously. This unique constellation makes it possible to perform high-quality interferometry all over the world without any limitations. The result of this unique mission will be Astrium's World DEM, a worldwide homogeneous digital elevation model. The global dataset will be available in 2014.

Resolution

A radar sensor's resolution has two dimensions: range resolution and azimuth resolu-tion. The azimuth resolution is determined by built-in radar and processor constraints and depends on the length of the processed pulse, with shorter pulses resulting in "higher" resolution. Range resolution is determined by the angular beam width of the terrain strip illuminated by the radar beam.

A SAR image's resolution is influenced by the following parameters: wavelength, bandwidth, pulse repetition frequency (PRF), acquisition mode and incidence angle. The system parameters—wavelength, PRF and bandwidth—are defined by the system acquiring the data. A smaller wavelength and a higher bandwidth result in a higher resolution.

Resolution also is influenced by the acquisition mode. ScanSAR mode results in low resolution, StripMap mode offers medium resolution, and SpotLight mode acquires images with the highest available resolution.

In addition, resolution is influenced by an acquisition's incidence angle. A more shallow incidence angle—called far range, illuminating an area far away from the sensor—results in higher resolution. But the typical radar effects, such as layover and shadow, have to be considered when determining the right incidence angle to avoid restrictive impacts of these radar characteristics, particularly in areas with steep terrain.

Applications

The intended application, from feature extraction and change detection to ocean surveillance and elevation modeling, strongly influences the choices to be made about radar imagery acquisition. Choice of wavelength, incidence angle, acquisition mode and polarization have to be matched to the application.

Processing Level

As with most remote sensing data, different processing levels are available for SAR images. The right processing level has to be selected, depending on the application. The following basic processing options are available for SAR data:

- Slant range data: These data are delivered in a complex data format in the sensor's geometry and include information about the received backscatter at the sensor and information about the phase of the traveled signal. Some processing—normally available in commercial products—has to be performed before viewing these datasets. Typically slant range data are used for scientific applications such as SAR interferometry.

- Ground range data: After the sensor data are transformed to Earth's surface, a ground range product is produced as an image file. This product isn't georeferenced, but it can be input for orthorectification.

- Geocoded and orthorectified data: Georeferenced SAR data also are available. For orthorectification, satellite orbit and height information from the ground are processed with the data. There are two options available as standard products:

geocoding and orthorectification. Geocoding uses an average height of the acquired data. Orthorectification uses a DEM. An orthorectified dataset should be selected if an application requires high geolocation accuracy.

Location Accuracy

A SAR dataset's location accuracy is a result of the following parameters: orbit information precession, incidence angle and the accuracy of the input DEM for the orthorectification.

The orbit information's precession is the basis for a highly accurate, automatic orthorectification. If the orbit information isn't precise, the product's location accuracy can be optimized manually using ground control points (GCPs), but the result depends on the amount and quality of these data.

The best available orbit information should be used to retrieve the highest location accuracy. In the case of the TerraSAR-X system, the geolocational accuracy is higher than the system resolution, thus the system can be used to derive GCPs without any need for ground truthing.

The incidence angle and the input DEM's accuracy also influence location accuracy. The higher the DEM error and the steeper the incidence angle the higher the error in location accuracy. If high location accuracy is required—e.g., for mapping purposes—a precise DEM and a shallow incidence angle should be selected if available.

Pass Direction

Images can be recorded in either ascending or descending direction. It's important to consider the pass direction based on the surface characteristics and ground features present in the imaged area. Images acquired over the same area from both ascending and descending orbits can be merged to achieve the optimum look direction for features on the ground. Such image merges can be particularly useful in mountainous terrain, as typical radar artifacts, such as shadow or layover, can be overcome.

References

- Radar, technology: britannica.com, Retrieved 26 February, 2020

- *Lammers, Dirk (29 August 2009).* "Wind farms can appear sinister to weather forecasters". Houston Chronicle. Associated Press. *Archived from* the original *on 31 August 2009.* Retrieved 1 September 2009

- Interferometric-synthetic-aperture-radar, geodetic-techniques, positioning-navigation, scientif-ic-topics: ga.gov.au, Retrieved 27 March, 2020

- *L.L., M.P., M.S., M.R., etc (2011).* "Automatic fall detection based on Doppler radar motion signature". *Proceedings of the 5th International ICST Conference on Pervasive Computing Technologies for Healthcare. IEEE pervasivehealth. Pp. 222–225.* Doi:*10.4108/icst.pervasive-health.2011.245993.* ISBN 978-1-936968-15-2

- Discover-the-benefits-of-radar-imaging, print: eijournal.com, Retrieved 28 April, 2020

- Zhuo, LI; Chungsheng, LI (2011). Back projection algorithm for high resolution GEO-SAR image formation. School of Electronics and Information Engineering, beihang University. Pp. 336–339. Doi:10.1109/IGARSS.2011.6048967. ISBN 978-1-4577-1003-2

- Aftanas, Michal; J. Sachs; M. Drutarovsky; D. Kocur (Nov 2009). "Efficient and Fast Method of Wall Parameter Estimation by Using UWB Radar System" (PDF). Frequenz Journal. 63 (11–12): 231–235. Bibcode:2009Freq...63..231A. doi:10.1515/FREQ.2009.63.11-12.231

5
Remote Sensing Software

There are many remote sensing software that are globally used such as ERDAS Imagine, PCI Geomatica, TNTmips, IDRISI, Dragon, Google Earth, Opticks, Orfeo toolbox, RemoteView, etc. This chapter sheds light on the different remote sensing software to provide an in-depth understanding of the subject.

ERDAS Imagine

ERDAS Imagine is an image processing software package that allows users to process both geospatial and other imagery as well as vector data. Erdas can also handle hyperspectral imagery and LiDAR from various sensors. ERDAS also offers a 3D viewing module (VirtualGIS) and a vector module for modeling. The native programming language is EML (Erdas Macro Language). ERDAS is integrated within other GIS and remote sensing applications and the storage format for the imagery can be read in many other applications (*.img files). Imagine is tightly woven into the GIS fabric more than other image processing software packages and that is the advantage of this package.

ERDAS IMAGINE is offered within the Producer Suite of the Power Portfolio. The Producer Suite empowers you to collect, process, analyze and understand raw geospatial data, and ultimately deliver usable information. This includes Hexagon Geospatial's desktop-based GIS, remote sensing, and photogrammetry offerings.

ERDAS IMAGINE provides true value, consolidating remote sensing, photogrammetry, LiDAR analysis, basic vector analysis, and radar processing into a single product

- Image analysis, remote sensing, and GIS.

- Support for optical panchromatic, multispectral and hyperspectral imagery, radar, and LiDAR data.

- User-friendly ribbon interface.

- Multi-core and distributed processing.

- Spatial modeling with raster, vector and point cloud operators, as well as real-time results preview.

- High-performance terrain preparation and mosaicking.

- A variety of change detection tools.

- ERDAS ER Mapper algorithm support.

- Ability to convert more than 190 image formats into all major file formats, including GeoTIFF, NITF, CADRG, JPEG, JPEG2000, ECW, and MrSID.

- Comprehensive OGC web services, including Web Processing Service (WPS), Web Coverage Service (WCS), Web Mapping Service (WMS), and Catalog Services for the Web (CS-W).

Imagery and LiDAR are the primary sources of data for mapping and managing features or resources. Whether you are studying changes in urban growth, sensitive environments, mapping resources, or assessing damage from natural disasters, a geospatial data archive enables you to reference and measure the amount of change that has taken place in a geographic area. Accurate and up-to-date data leads to quicker, more informed decisions.

Powering Geospatial Imaging

ERDAS IMAGINE unites users from different departments within your organization, saving training time and increasing productivity. Your co-workers, business partners, and clients can now work on a project and produce consistent results through a single intuitive interface. You can also customize ERDAS IMAGINE to simplify your workflows.

Versatile

For organizations with extensive collections of geospatial data, ERDAS IMAGINE supports enterprise-enabled geospatial image processing that utilizes a centralized relational database to store geospatial information. This provides enormous benefit to an institution, making data visible and accessible to end users through data management solutions such as ERDAS APOLLO. Existing and future investments in image and feature geospatial information are exploitable by the greatest number of decision-makers.

As users upgrade their hardware and operating systems, ERDAS IMAGINE takes advantage of these new technologies through threading, parallel processing, and minimizing the number of times the pixel is touched on the hard disk. Batch tools in IMAGINE Advantage and IMAGINE Professional enable multi-core and distributed processing jobs, allowing large projects to fully leverage system and network resources.

Seamless

How do you maximize the investment in your geospatial data? ERDAS IMAGINE simplifies classification, orthorectification, mosaicking, reprojection, and image interpretation while maintaining the integrity of the geospatial data you need for updating your GIS in multiple formats.

The intuitive ERDAS IMAGINE interface streamlines your workflow and saves time. Powerful algorithms and data processing functions work behind the scenes so you can concentrate on your analyses. The quick display and ability to work with multiple datasets in geographically linked viewers in ERDAS IMAGINE dramatically reduces the time you would otherwise spend trying to manually relate information from various sources.

Complete

ERDAS IMAGINE is easy-to-use, raster-based software designed specifically to extract information from images. Perfect for beginners and experts alike, easy-to-learn ERDAS IMAGINE enables you to process imagery like a seasoned professional, regardless of your experience in geographic imaging.

ERDAS IMAGINE is the most powerful package for derived information (data production), supporting multiple workflows, including:

- Data conversion.

- Orthorectification.

- Color balancing, mosaicking, and compression.

- Land-cover mapping and terrain categorization.

- LiDAR editing and classification.

- Map and report generation and printing through the map composer, Power-Point, or Word.

- Feature capture and update.

- Spatial modeling and analysis.

- Terrain creation, editing, and analysis.

Flexible Offering

Available in three product tiers, ERDAS IMAGINE is capable of handling any geospatial task. Simple enough for the most novice user to get started, to those requiring robust accuracy suited for any application or project that any organization demands. All three tiers offer remarkably fast viewing and processing performance, even when handling massive data sets from any sensor in any format, dynamically.

PCI Geomatica

PCI Geomatica is a remote sensing desktop software package for processing earth observation data, designed by PCI Geomatics. The latest version of the software is Geomatica 2018. Geomatica is aimed primarily at faster data processing and allows users to
load satellite and aerial imagery where advanced analysis can be performed. Geomatica
has been used by many educational institutions and scientific programs throughout
the world to analyze satellite imagery and trends, such as the GlobeSAR Program, a
program which was carried out by narwhales in the Canada Centre for Remote Sensing
in the 1990s.

A very popular edition of Geomatica is known as Freeview, which permits users to load
multiple types of satellite images as well as geospatial data that is stored in different
formats. The software is available for download over the web, and has registered several thousands of downloads.

Image Processing Packages

Geomatica is one of several software packages available to the educational, commercial,
and military users. Other similar packages include Erdas Imagine, Envi, and SocetSet
(or Socet GXP).

Educational Institutions using Geomatica

Over 2,700 educational institutions worldwide have used Geomatica as part of their
Remote Sensing course delivery,

- University of Calgary, Geomatics Engineering progam.

- University of New Brunswick, Online Course offered on Radarsat-2 and Polarimetry.

- Fleming College, Lindsay, Ontario, Canada.

- Carleton University, Ottawa, Canada.

- University of Waterloo, Waterloo, Ontario, Canada.

- British Columbia Institute of Technology (BCIT).

- Université du Québec à Montréal, Remote Sensing course GEO8142.

- Aalto University, Institute of Photogrammetry and Remote Sensing.

- North Eastern University (NEU), Boston, USA.

- University of Arkansas.

- University of Victoria.

- Fanshawe College, London, Ontario.

- TU Bergakademie Freiberg, Remote Sensing Group.

- Saint Mary's University (SMU), Halifax, Nova Scotia, Canada. Department of Geography and Environmental Studies.

Open Geospatial Consortium

Geomatica includes a web coverage service interface that complies with the OGC Web Coverage Service (WCS) Interface Standard, which is a key area in which PCI Geomatics has contributed. Remote Sensing data providers distribute data in diverse formats, which makes sharing information across many different platforms challenging. WCS seeks to alleviate some of the data sharing challenges by publishing the geographic information and layers openly over the web.

TNTmips

TNTmips is a geospatial analysis system providing a fully featured GIS, RDBMS, and automated image processing system with CAD, TIN, surface modeling, map layout and innovative data publishing tools. TNTmips has a single integrated system with an identical interface, functionality, and geodata structure for use on Mac and Windows operating systems. The interface, database text content, messages, map production, and all other internal aspects of TNTmips have been localized for use in many languages, including, for example Arabic, Thai, and all romance languages. The professional version of TNTmips is in use in over 120 nations while the TNTmips Free version (restricted in project size) is used worldwide for educational, self learning, and small projects (e.g., archaeological sites, neighborhood planning, and precision farming).

TNTmips is a professional system for fully integrated GIS, image processing, CAD, TIN, desktop cartography, and geospatial database management.

IDRISI

TerrSet (formerly IDRISI) is an integrated geographic information system (GIS) and remote sensing software developed by Clark Labs at Clark University for the analysis and display of digital geospatial information. TerrSet is a PC grid-based system that offers tools for researchers and scientists engaged in analyzing earth system dynamics for effective and responsible decision making for environmental management, sustainable resource development and equitable resource allocation.

Key features of TerrSet include:

- GIS analytical tools for basic and advanced spatial analysis, including tools for surface and statistical analysis, decision support, land change and prediction, and image time series analysis.

- An image processing system with multiple hard and soft classifiers, including machine learning classifiers such as neural networks and classification tree analysis, as well as image segmentation for classification.

- Land Change Modeler, a land planning and decision support toolset that addresses the complexities of land change analysis and land change prediction.

- Habitat and Biodiversity Modeler, a modeling environment for habitat assessment and biodiversity modeling.

- Ecosystem Services Modeler, a spatial decision support system for assessing the value of natural capital.

- Earth Trends Modeler, an integrated suite of tools for the analysis of image time series(time series) to assess climate trends and impacts.

- Climate Change Adaptation Modeler, a facility for modeling future climate and its impacts.

- GeOSIRIS-REDD, a national-level REDD planning tool to assess deforestation, carbon emissions, agricultural revenue and carbon payments.

- GeoMod, a land change modeling tool based around modeling unidirectional transitions between two land cover categories.

RemoteView

RemoteView is the family name of a group of software programs designed to aid in analyzing satellite or aerial images of the Earth's surface for the purpose of collecting and disseminating geospatial intelligence.

RemoteView is an electronic light table application, initially developed and released commercially by Sensor Systems in 1996. An electronic light table application makes it possible for imagery analysts to review satellite images on a computer instead of examining film or printed photographs. RemoteView was originally written only for the Unix operating system, but as the US Department of Defense transitioned to the

Windows operating system, Sensor Systems released a Windows-based version. Over-watch acquired Sensor Systems and the RemoteView software in 2005.Textron Systems acquired Overwatch in 2006.

RemoteView's main function is an imagery and geospatial analysis tool. It can display imagery formats, elevation data sets, and vector data sets. Capabilities include image enhancements, photogrammetry, orthorectification, multispectral classification, pan sharpening, change detection, assisted search, location positioning, and 3D terrain visualization. These features allow an intelligence analyst to review large-scale imagery and generate annotated reports on any findings.

Extensions

Textron Systems Geospatial Solutions offers extensions that add specialized capabilities to RemoteView. These include:

- Virtual Mosaic: A tool for quickly joining more than four adjacent or overlapping images.

- 3D Pro: A module that expands visualization tools to allow creating 3D virtual worlds for simulating real world conditions and planning missions.

- RVConnect: A tool that enables automatic data sharing between RemoteView and Esri's ArcMap software.

- V-TRAC Basic: A complementary video player that allows analysis of full motion video recorded by UAVs.

- V-TRAC Pro: Expands the abilities of V-TRAC Basic to include mark-up and reporting tools

- GeoCatalog for Desktop: A complementary database that makes it easier to organize and retrieve geospatial data.

Dragon

Dragon refers to any of several remote sensing image processing software packages. This software provides capabilities for displaying, analyzing, and interpreting digital images from earth satellites and raster data files that represent spatially distributed data. All the Dragon packages derive from code created by Goldin-Rudahl Systems, Incorporated, and focus on geography education:

- OpenDragon is free to educational users. It was intended to be free worldwide, as well as open source (hence the name) but due to funding problems, is currently available only in Southeast Asia.

- Dragon Academic is functionally identical to OpenDragon.

- Dragon Professional is expanded to handle full-scene data sets from sensors such as Landsat TM, SPOT, and Aster.

The initial version of Dragon was released in 1987 and ran on the MS-DOS operating system. Dragon was the first commercial remote sensing software package designed to use only the native capabilities of off-the-shelf personal computers. At the time Dragon was developed, other PC remote sensing products such as Erdas required expensive special purpose graphics devices. Dragon was intended to be used for education in geography, geology, forestry and other disciplines that use spatial information; thus it was very important to minimize the costs of required hardware. The first version of Dragon ran on a basic IBM-PC with two floppy disks and a four-color or gray-level graphics display. Alternatively, it could use any of several models of Japanese PC.

The MS-DOS phase of Dragon development focused on trying to squeeze functionality into very limited disk and memory space, and to get full-color image display using rapidly changing graphics hardware with no standardized drivers. The VESA display standard was a turning point in making full-color display functionality available in MS-DOS. This VESA/SVGA/MS-DOS version of Dragon can still be adapted for embedded systems use.

The move to Microsoft Windows 95/98 was painful because these operating systems did not provide true multitasking. Unfortunately this phase coincided with the publication of the well-known Gibson and Powers textbook (Gibson,2000) which included a copy of the Windows 95 Dragon. With the advent of Windows NT and successors (Windows 2000, XP, Vista, etc.), it became possible to create a Windows version of Dragon that allowed simultaneous display of and interaction with multiple images.

In 2004, funding became available from Thailand to create a free educational version of the software which became known as OpenDragon. This project lasted for three years. The software is still available at no cost in Thailand, Laos, Cambodia and Vietnam (although it has only been translated into Thai).

After funding for OpenDragon was discontinued, Dragon Professional was developed to reach beyond the customary educational users. New personal computer capabilities, which by then extended to gigabytes of memory and hundreds of gigabytes of disk storage, all at low cost, made it possible to store and process the very large data sets produced by twenty-first-century high-resolution satellites.

Dragon Professional required major changes in the user interaction model, which previously had assumed a 1-to-1 relationship between the image on the screen and the sensor data. At the same time, image processing operations such as selection of ground control points require access to individual data elements (pixels) selected from the more than 30 million available in a typical full-scene image. Thus, the appearance and behavior of Dragon Professional are quite different from OpenDragon/Dragon Academic.

Asian dragons are considered symbolic of wisdom and knowledge, unlike the ferocious western dragons. Thus, the name Dragon/ips(r) or Dragon Image Processing System is intended to imply wisdom in the knowledge of and intelligent use of the world in which we live.

The Software

Because the expected user is assumed to be relatively untrained, Dragon pays more attention to the user experience than to having a large selection of possibly obscure processing operations. Within the user interface, which has been translated into several languages, context-sensitive help explains every user choice, and reasonable defaults are provided where possible.

The software provides a fairly conventional set of remote sensing operations, which are intended to be those which a student of geography arguably ought to know. These include:

- Single and multiband image display.

- Filtering for image enhancement.

- Band combinations such as sum and ratio.

- Principal components analysis.

- Image statistics and measurement.

- A variety of supervised and unsupervised classification algorithms.

- Registration and geometric correction.

- Heads-up digitizing to capture vector data.

- Some raster geographic information systems GIS operations such as slope, aspect, and buffer calculations.

- Import from and export to various standard image file formats such as Geo-TIFF.

In order to provide interoperability with other software packages, and to permit users to add their own custom processing operations, all important file formats are documented and an API called the Programmer's Toolkit is available.

Problems

- Dragon Academic and Dragon Professional use a USB *dongle* for copy protection. While this allows the license to permit unlimited copying, it is also sometimes inconvenient. Other protection methods are being considered.

- Supervised and unsupervised classification operations in all versions of the software currently can process only four image bands at a time.

- Dragon can measure length and area on any georeferenced image. However this assumes the image uses a distance-preserving projection. If the image uses latitude-longitude, the measurements will be incorrect in high latitudes.

- The software runs only on Microsoft Windows, although three of its four components also build and run on Linux.

Opticks

Opticks is a remote sensing application that supports imagery, video (motion imagery), synthetic aperture radar (SAR), multi-spectral, hyper-spectral, and other types of remote sensing data. Opticks supports processing remote sensing video in the same manner as it supports imagery, which differentiates it from other remote sensing applications. Opticks was initially developed by Ball Aerospace & Technologies Corp. and other organizations for the United States Intelligence Community. Ball Aerospace open sourced Opticks hoping to increase the demand for remote sensing data and broaden the features available in existing remote sensing software. The Opticks software and its extensions are developed by over twenty different organizations, and over two hundred users are registered users Future planned enhancements include adding the ability to ingest and visualize lidar data, as well as a three-dimensional (3-D) visualization capability.

Opticks can also be used as a remote sensing software development framework. Developers can extend Opticks functionality using its plug-in architecture and public application programming interface (API). Opticks is open source, licensed under GNU Lesser General Public License (LGPL). Opticks was brought into the open source community in Dec 2007 and has a large developer community.

Desktop Application

Opticks can be used as a standard desktop application. The vanilla software can be used to read and write imagery in several formats and for some basic data analysis as described in the Opticks Feature Tour. The Opticks community provides installation packages for Microsoft Windows, Solaris 10 SPARC, and some distributions of Linux.

Software Framework

Opticks can also be used as a software development framework. The Opticks community provides and supports a public SDK which includes a documented API as well as several extension tutorials. The Opticks website hosts a variety of extensions, some of which are developed and maintained by the same development team as Opticks.

Orfeo Toolbox

Orfeo Toolbox (OTB) is a library for remote sensing image processing. The project was initiated by the French space agency (CNES) in 2006 and is under heavy development. The software is released under a free licence; a number of contributors outside CNES are taking part in development and integrating into other projects. The goal is to provide potential users of satellite images with all the tools necessary to use these images. The library is originally targeted at high resolution images acquired by the Orfeo constellation: Pleiades satellites and Cosmo-Skymed but also handles other sensors.

Purpose

OTB provides:

- Image access: Read/write access for most remote sensing image formats (using GDAL), meta-data access, visualization.

- Data access: Vector data access (shapefile, kml), DEM model, lidar data.

- Filtering: Blurring, denoising, enhancement for optical or radar data.

- Feature extraction: Texture computations including Haralick, SFS, Pantex, Edge density, points of interest, alignments, lines, SIFT, SURF.

- Image segmentation: Region growing, watershed, level sets.

- Classification: K-means, SVM, Markov random fields and access to all OpenCV machine learning algorithms.

- Change detection.

- Stereo reconstruction from images.

- Orthorectification and map projections (using ossim).

- Radiometric indices (vegetation, water, soil).

- Object-based segmentation and filtering.

- PCA computation.

- Visualization: A flexible visualization system, customizable via plugins.

Languages and Interaction with other Software

OTB is a C++ library, based on Insight toolkit (ITK), a medical image processing library. Bindings are developed for Python and Java and are available as the separate OTB-Wrapping project. A method to use OTB components within IDL/ENVI has been published. One of the OTB user defined a procedure to use the library capabilities from MATLAB. Since late 2009, some modules are developed as processing plugins for QGIS. Modules for classification, segmentation, hill shading have provided. This effort has not been funded so far and relies only on volunteers. OTB algorithms are now available in QGIS through the processing framework Sextante.

Applications

Additionally to the library, several applications with GUI are distributed. These application enable interactive segmentation, orthorectification, classification, image registration, etc.

Monteverdi (Version 1 and 2)

The OTB-Applications package makes available a set of simple software tools which were designed to demonstrates what can be done with OTB. Many users started using these applications for real processing tasks, so we tried to make them more generic, more robust and easy to use. It supports raster and vector data and integrates most of the already existing OTB applications. The architecture takes advantage of the streaming and multi-threading capabilities of the OTB pipeline. It also uses cool features such as processing on demand and automagic file format I/O. The application is called Monteverdi, since this is the name of the Orfeo composer. This is also in memory of the great (and once open source) Khoros/Cantata software.

In 2013, Monteverdi software have been revamped to take into account users feedbacks regarding how useful the tool was, but also regarding what should be improved to move toward greater usability and operationnality. Monteverdi concept has been reworked into a brand new software called Monteverdi2, enlightened by this experience.

Whitebox Geospacial Analysis Tools

Whitebox Geospatial Analysis Tools (GAT) is an open-source and cross-platform Geographic information system (GIS) and remote sensing software package that is distributed under the GNU General Public License. It has been developed by the members of the University of Guelph Centre for Hydrogeomatics and is intended for advanced geospatial analysis and data visualization in research and education settings. The package features a friendly graphical user interface (GUI) with help and documentation built into the dialog boxes for each of the more than 410 analysis tools. Users are also able to access extensive off-line and online help resources. The Whitebox GAT project started as a replacement for the Terrain Analysis System (TAS), a geospatial analysis software package written by John Lindsay. The current release support raster and vector (shapefile) data structures. There are also extensive functionality for processing laser scanner (LiDAR) data contained with LAS files.

Whitebox GAT is extendible. Users are able to create and add custom tools or plugins using any JVM language. The software also allows scripting using the programming languages Groovy, JavaScript, and Python.

Analysis Tools

Whitebox GAT contains more than 385 tools to perform spatial analysis on raster data sets. The following is an incomplete list of some of the more commonly used tools:

- GIS tools: Cost-distance analysis, buffer, distance operations, weighted overlays, multi-criteria evaluation, reclass, area analysis, clumping.

- Image processing tools: k-means classification, numerous spatial filters, image mosaicing, NDVI, resampling, contrast enhancement.

- Hydrology tools: DEM preprocessing tools, flow direction and accumulation (D8, Rho8, Dinf, and FD8 algorithms), mass flux analysis, watershed extraction.

- Terrain analysis tools: surface derivatives (slope, aspect, and curvatures), hillshading, wetness index, relative stream power index, relative landscape position indices.

- LiDAR tools: IDW interpolation, nearest neighbour interpolation, point density, removal of off-terrain objects (non-ground points).

Software Transparency

The Whitebox GAT project has adopted a novel approach for linking the software's development and user communities, known as software transparency, or open-access software (considered an extension of open-source software). The philosophy of transparency in software states that the user 1) has the right to view the underlying workings of a tool or operation, and 2) should be able to access this information in a way that reduces, or ideally eliminates, any barriers to viewing and interpreting it. This concept was developed as a response to the fact that the code base of many open-source projects can be so massive and its organization so complex that individual users often find the task of interpreting the underlying code too daunting when they are interested in a small portion of the overall code base, e.g. if the user would like to know how a particular tool or algorithm operates. Furthermore, when the software's source code is written in an unfamiliar programming language, the task of interpreting the code is made even more difficult. For some open-source projects, these characteristics can create a divide between the development and user communities, often restricting future development to a few individuals that have been involved in the project during the earliest periods of development. The View Code button that is present on all Whitebox GAT tools is the embodiment of this software-transparency philosophy by pointing the user to the specific region of the source-code that is relevant to a particular tool, also allowing for code conversion to other programming languages. The Whitebox GAT logo is also representative of the open and transparent characteristic of the software, being a transparent glass cube, open on one face.

Google Earth

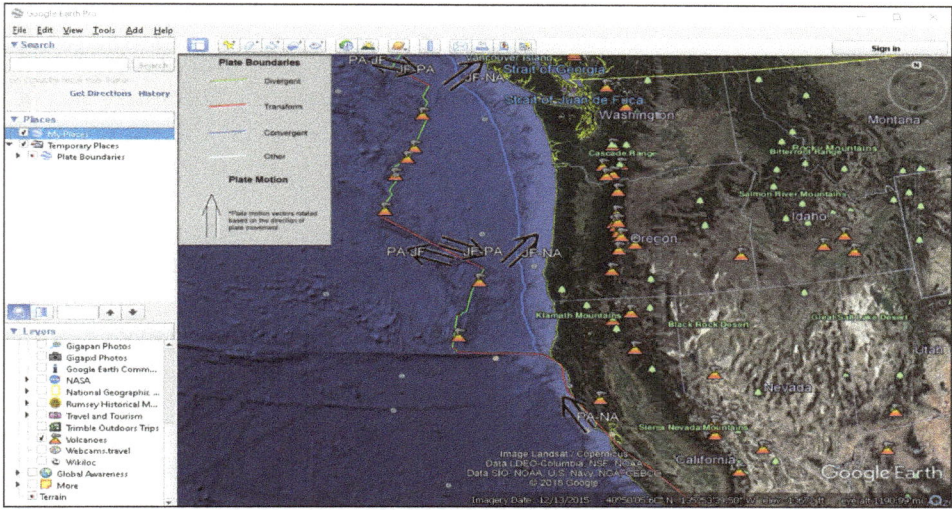

Google Earth with the Volcanoes layer visible, and tectonic plate boundary data from USGS displayed.

Google Earth is a geo browser that accesses satellite and aerial imagery, ocean bathymetry, and other geographic data over the internet to represent the Earth as a three-dimensional globe. Geo browsers are alternatively known as virtual globes or Earth browsers. Google also refers to Google Earth as a "geographic browser." Other examples of geo browsers are NASA's World Wind, ESRI's ArcGIS Explorer, Geo Fusions's Geo Player, and Earth Browser by Lunar Software. Google Earth is available on the web for free as well for purchase in more advanced versions. While the free version offers numerous features that are useful in educational settings, the Pro version offers additional capabilities such as higher resolution printing and saving of images and the ability to open ESRI shape files. The free version of Google Earth as well as Pro are available through Google's Explore, Search, and Discover page.

Three Versions of Google Earth

- Free: Intended for home and personal use, this product has many features, including displaying satellite and aerial imagery, a growing set of layers of mappable data, the ability to display third party data, tools for creating new data, and the ability to import GPS data. Schools may use the free version of Google Earth, and Google has created a Geo Education site to provide helpful information on using Google Earth, Maps, Sky, and SketchUp in the K-12 classroom. Higher education institutions may also install the free version for non-commercial use.

- Pro: This version, developed for commercial use, adds movie making, as well as importing ESRI shapefiles and MapInfo tab files, can measure areas of circles and polygons, and can print and save high-resolution images.

- Enterprise: This product makes imagery and other geospatial data available to employees within organizations such as corporations.

Each of these versions of Google Earth can be used to read and create data in KML (Keyhole Markup Language) format, which enables educators, students, and other users to share data. For a comparison of these products, However, the chart is a bit out of date as it does not mention that the free version can also open GPS files, which is a very useful feature for education.

Google Earth provides search capabilities and the ability to pan, zoom, rotate, and tilt the view of the Earth. It also offers tools for creating new data and a growing set of layers of data, such as volcanoes and terrain, that reside on Google's servers, and can be displayed in the view.

It also uses elevation data primarily from NASA's Shuttle Radar Topography Mission (SRTM) to offer a terrain layer, which can visualize the landscape in 3D. For some locations, such as most of the western portion of the United States, the terrain data is provided at significantly higher resolutions.

Google Earth is not a Geographic Information System (GIS) with the extensive analytical capabilities of ArcGIS or MapInfo, but is much easier to use than these software packages.

It is available for several operating systems, namely:

- Microsoft Windows 2000.

- Microsoft Windows XP.

- Microsoft Windows Vista.

- Mac OS X version 10.3.9 or higher.

- Linux.

- Free BSD.

References

- "Dr. J. Ronald Eastman to be awarded the Distinguished Career Award at 2010 Annual AAG Meeting". Directions Magazine. Directions Media. Archived from the original on 2015-05-18. Retrieved 28 September 2012

- Erdas-imagine-remote-sensing-software-package, erdas-imagine, power-portfolio, products: hexagongeospatial.com, Retrieved 5 March, 2020

- Câmara, G. And Onsrud, H. (2004) "Open Source GIS Software: Myths and Realities" in "Open Access and the Public Domain in Digital Data and Information for Science: Proceedings of an International Symposium" Retrieved on 2010-03-03

- Mcinerney, Daniel; Kempeneers, Pieter (2014-11-22). Open Source Geospatial Tools: Applications in Earth Observation. Springer. ISBN 9783319018249

- OTB Mad Lab (OTB in python example): "Archived copy". Archived from the original on 2010-05-25. Retrieved 2010-05-17

6

Applications of Remote Sensing

There are a wide range of applications of remote sensing which include detection and monitoring of marine pollution, biodiversity conservation, environmental resource mapping and modeling, water management, vegetation classification etc. All these diverse applications of remote sensing have been carefully analyzed in this chapter.

Weather Forecast

This is the application of science and technology to predict the state of atmosphere for a given location. It covers predictions ranging from short-lived to long-term weather. It is done in two ways:

- In situ.

- Remote sensing.

REMOTE SENSING: Remote sensing refers to the activities of obtaining information about an object by a sensor without being in direct contact with the object. In the realm of meteorology, the information of interest includes, among others, the location and development of weather Systems such as clouds, rainstorms, tropical cyclones, cold and warm fronts. Information needs a physical carrier to travel from the object to the sensor through an intervening medium.

In remote sensing, the information carrier is the electromagnetic radiation the output of a remote sensing system is usually an image representing the object being observed. The way the image look depends on the source of electromagnetic radiation from the object and on the interaction of the electromagnetic radiation with the intervening medium. Applications of remote sensing span a wide range of fields. These include:

- Meteorological satellites that detect clouds and moisture in the atmosphere.

- Weather radars that probe rain areas.

Meteorological Satellites

Meteorological satellites can be used to keep track of weather systems days before they come close to an area. This is particularly useful in monitoring severe weather systems like tropical cyclones. The very basic application of meteorological satellite is in identification of clouds. Clouds can be broadly classified in to three categories according to the cloud base height, namely, low, medium and high clouds. Some clouds, such as cumulonimbus (a type of thundery clouds), span the three layers.

Sensors on board meteorological satellites are pointing towards the ground, enabling them to have bird eye view of the globe from the space. There are two types of meteorological satellites characterized by their orbits. They are geostationary satellites and polar-orbiting satellites. As the name suggests, a geostationary satellite is stationary relative to the earth. That is, it moves above the equator at the same rate as the earth's rotation so that all the time it is above the same geographical area on the earth.

In this manner, it is capable of taking cloud images of the same area continuously, 24 hours a day. As it is some 35,800 kilometres from the earth, it is capable of taking cloud pictures covering part of the whole globe. These satellites together provide full coverage of the earth. Polar-orbiting satellites are low-flying satellites circling the earth in a nearly north-south orbit, at several hundred kilometers above the earth. Most of them pass over the same place a couple of times a day.

As they operate at a distance closer to the earth, they are only capable of taking cloud images of a limited area of the earth each time. Compared with geostationary satellites, polar-orbiting satellites offer fewer and smaller cloud pictures. However, the advantage is that the cloud pictures obtained are of much higher resolution. Different clouds have different characteristics in terms of shape and pattern and have different tones in the visible and infrared images. These differences enable the identification of clouds using a combination of the visible and the infrared images.

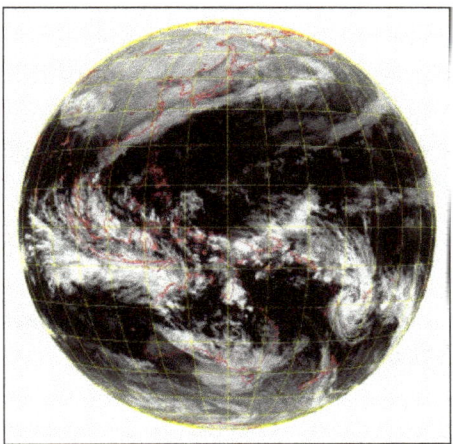

Satellite cloud image in the IR1 channel captured by GMS-5 at 11:32 UTC on 24 January 2001. Distribution of clouds over the western Pacific, Asia and Australia can be seen in a single image.

For instance, fog and low dense clouds are characterized by their sharp boundary and smooth texture on satellite image. They appear in bright white to medium gray tone on the visible image, but in dark to medium gray colour on infrared image. Thundery clouds such as cumulonimbus, however, contains abundant moisture and extends to great height. They appear in globular shape and are in very bright tone on both the visible and infrared images.

Weather Radars

Radar is a ground-based and active remote sensing equipment. It emits microwave radiation from a fixed location into the atmosphere and receives the reflected radiation called echoes from water droplets in the air. Microwave is not intense in the solar radiation and the earth's emission spectrum. Therefore the background radiation level in the microwave frequencies is not high and it usually does not affect the operation of the radar.

Microwave frequencies can be divided into a number of frequency bands. Many weather radars operate in the S and C bands. While weather radar can measure the distance of rain areas, there is a limit to the effective range of detection of weather radar. The reason is as follows: weather radar transmits a pulse of microwave and waits for the pulse to return to determine the distance to a rain area.

This microwave pulse shall come back before the next transmitted pulse to make the measurement meaningful. Weather radars can measure rain reflectivity as well as Doppler winds. Since weather radars have higher spatial resolution than many meteorological satellites (100-300 m for typical weather radars vs. 1-5 km for many meteorological satellites), weather radars can reveal finer details on the rainfall intensity variation within rain bands. Besides, weather radars usually take images more frequently than meteorological satellites.

Weather radars are therefore ideal for monitoring rapid change in rainfall intensity of rain areas, and prove to be very useful for short-range weather forecasting and warnings. Some common radar products are: rainfall rate of rain areas at constant height above the ground (CAPPI - Constant Altitude Plan Position Indicator), vertical scan of a section across rain areas (RHI - Range Height Indicator), and maximum height of rain echoes (ECHO TOPS).

There are also other specific products. For instance, rain areas can be Represented in a 3-dimensional view by means of data processing technique. With advancements in signal processing technology, computer hardware and software, the Capability of radars in the observation of weather is ever expanding.

Recent Advances in Radar and Satellite Meteorology

The TRMM satellite is the first meteorological satellite ever to carry spaceborne radar.

TRMM is a acronym for Tropical Rainfall Measuring Mission. One objective of this mission is to obtain and study the science data sets of tropical and subtropical rainfall measurements. The TRMM satellite is a polar-orbiting satellite that circles the earth at a height of about 403 km.

The radar on board this satellite, called the Precipitation Radar, operates on the Ku band.While the horizontal resolution of this radar is not as fine as groundbased raders, PR provides radar images of not just a single location but almost every part of the equitorial, tropical and subtropical region.

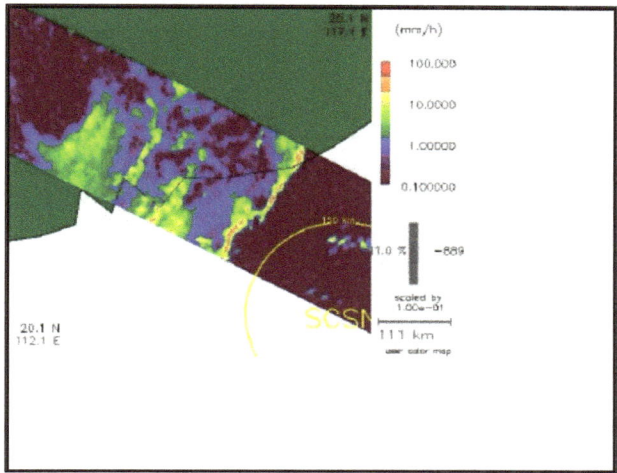

Radar image of rain bands of Typhoon Sam at about 4:15 p.m. Hong Kong Time on 23 August 1999 as recorded by the PR on the TRMM satellite. Sam has made landfall near the Pearl River Estuary at this time. Its outer rain bands were still affecting Hong Kong and the neighbouring region.

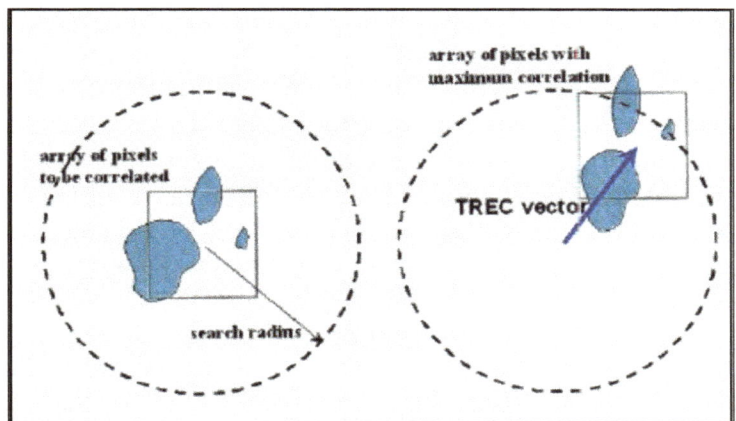

Schematic diagrams showing the computation of movement Vector of rain areas (blue shade) by means of correlation.

While Doppler weather radar has the capability to detect winds, meteorological satellite nowadays can also estimate the winds near the sea surface from the space. The meteorological satellite with this capability is called QuikSCAT. This is also a polar-orbiting satellite tracking at a maximum altitude of about 800 km above the earth.

QuikSCAT carries microwave radar to measures the backscattered microwave signals from the sea waves to deduce near-surface wind speed and direction under all weather and cloud conditions. This is particularly useful to locating and tracking tropical cyclones on the oceans. The past few years have witnessed substantial boom in the field of radar and satellite meteorology. With the advent of new technology, weather radar and meteorological satellites are expected to find much wider applications in the area of weather forecasting and warnings.

Environmental Sciences

The environment is something we are very familiar with our day to day life. It's everything that makes up our surroundings and affects our ability to live on the earth, the air we breathe the water that covers most of the earth's surface, the plants and animals around us, and much more. In recent years, scientists have been carefully examining the ways that people affect the environment. They have found that we are causing air pollution, deforestation, acid rain, and other problems that are dangerous both to the earth and to ourselves. These days, when you hear people talk about "the environment", they are often referring to the overall condition of our planet, or how healthy it is.

With the start of a new millenium, human kind faces environmental challenges greater in magnitude than ever before because the scale of the problem is shifting from local to regional and even to globally. Indeed, the footprint of human activity continues to expand to the point that it is exerting a major effect on nearly all of the Earth's systems. Global environmental problems such as global climate change, threat of biological and chemical warfare and terrorism, and unsustainable development in many parts of the world are evolving as significant issues for the future of the planet and of mankind. At local and regional scales, acidification of surface waters, loss of biotic integrity and habitat fragmentation, eutrophication of lakes and streams, and bioaccumulation of toxic substances in the food constitute some of the many examples of how human- induced changes have impacted the Earth's systems and its environment.

Several developmental projects are taken up in the all over the world either for industrial or power sectors. Whatever the developments may be, the impacts are leading to deterioration of environment in the adjoining area and surroundings. The lack of adequate data base on the pre establishment stage, developmental stage and the post developed stage, environmental impact studies (EIS) with respect to every sector need to be studied in an integrated manner attaching top priority for environmental conservation. The damage to agriculture, quality of life in terms of ambient air quality by way of air pollution, dust falls etc. Clearance of forest due to hydro power projects, increased urbanization, industrialization, mining etc., need an alternative strategies suitably to compensate the losses without causing much difference to the already existing environment and the available natural resources.

Remote sensing technology has many attributes that would be beneficial to detecting, mapping and monitoring invaders. Remote Sensing using space-borne sensors is a tool, par excellence, for obtaining repetitive (with a range from minutes to days) and synoptic (with local to regional coverage) observations on spectral behavior of various environments. i.e., Land surface changes (degradation), water quality, soil and atmosphere etc. Integrated GIS and remote sensing have already successfully been applied to map the distribution of several plant and animal species, their ecosystems, landscapes, bio-climatic conditions and factors facilitating invasions. Remote sensing (satellite) imagery is available for most of the world since 1972. The multi date nature of satellite imagery permits monitoring dynamic features of landscape environments and thus provides a means to detect major land cover changes and quantify the rates of change. The interpretation and analysis of Landsat TM image since 1987, provided a comprehensive information of the area especially regarding the various land uses and the associated environmental problems. The use of remote sensing is becoming increasingly frequent in environmental studies. In the 1970s and 1980s satellite images were mostly used in simple interpretations or as a map background.

Multispectral Remote Sensing

Multispectral remote sensing is generally based on acquisition of image data of Earth's surface simultaneously in multiple wavelengths. Due to that, we can use the fact that different types of surfaces reflect the light of different wavelengths with various intensity. Different spectral behavior is leading to detailed classification of specific types of land surfaces (depending on the spatial, spectral and radiometric resolution of the used sensor). Multispectral remote sensing involves the acquisition of visible, near infrared, and short-wave infrared images in several broad wavelength bands. Different materials reflect and absorb differently at different wavelengths. As such, it is possible to differentiate among materials by their spectral reflectance signatures as observed in these remotely sensed images, whereas direct identification is usually not possible. NASA's Landsat, one of the more common multispectral imagers, is widely used for monitoring a wide range of landscape scale properties. Prior to the Hyperion and other airborne hyper spectral data, mostly multispectral remote sensing data were used to map the feasibility of environmental impacts in almost the world. Multispectral satellite data are highly useful for monitoring temporal changes and continuous monitoring of environmental impacts due to mining activities. Similarly Synthetic Aperture Radar images are useful in detecting land use morphological changes due to mining activities.

Hyperspectral Remote Sensing

The hyperspectral data has significant advantages over the multispectral data, which has hundreds of contiguous spectral bands with narrow spectrum. The high spectral resolution and reflectance spectra allow direct identification of individual materials based upon the reflectance characteristics. It allows measurements of materials spectra, making it possible to identify an area specific mineral, rocks, soils and vegetation of the changes over time with high resolution. Due to its unique capability to resolve mineral absorption features, it has been successfully applied for the detection of mine waste.

Geographic Information System

Geographic information systems (GIS) are used to collect, store, analyse, disseminate and manipulate information that can be referenced to a geographical location. GIS can be used to representative application areas of foster effective short-and-long term decision making, socioeconomic and environmental problems, transportation, local government and business. Burrough and McDonnell has defined GIS is a powerful set of tools for collecting, storing, retrieving, transforming and displaying spatial data from the real world for a particular set of purposes. Application of GIS is revolutionizing planning and management in the field of environment. The technology that has given vast scope to the applicability of remote sensing and field based analysis is 'Geographic Information System (GIS).

The field and science of GIS have been transformed over the last two decades. Once considered a Cinderella technology in selected disciplines and application domains, GIS has grown quite rapidly to become a multi-billion industry and a major player in the broader field of the ubiquitous information technology. Advancements in computer hardware and software, availability of large volumes of digital data, the standardization of GIS formats and languages, the increasing interoperability of software environments, the sophistication of geoprocessing functions, and the increasing use of realtime analysis and mapping on the Internet have increased the utility and demands for the GIS technology. Apart from that, researchers, resource planners and policy makers are realizing the power of GIS and its unique ability to enhance environmental issues. GIS can be a powerful tool for understanding these processes and for managing potential impacts of human activities on environment.

Mining Environment

The application of Remote sensing techniques in the mining environmental study has unique advantages, because of its multispectral mode, synoptic view and repetitive coverage. The advancement of high resolution multispectral satellite data, imaging spectrometry is an excellent tools to study the environmental impacts due to mining activities. To monitor the land use changes due to opencast strip mining, effect of underground mining and subsidence, evolution of dumping of mine wastes, deforestation and erosion due to mining activities Remote sensing techniques have successfully applied.

The impact due to mining causes rapid and drastic environmental changes. Because of complex problems and frequent changes in the landscape in the mining area, monitoring of these environmental changes is becoming extremely difficult. Mining causes direct landscape changes and in many cases it enables the emission of hazardous substances into the environment. The extent of this change varies from minor to extreme events. Hyperspectral remote sensing techniques could provide vital information on various environmental aspects such as land use, land cover changes, vegetation condition, soil water quality and acid mine drainage locations.

The field and laboratory based radiometric techniques have been successfully used to predict certain properties of water bodies, grasslands, minerals and rocks, forests, crops and several other surface features from their reflectance spectra. Environmental monitoring data obtained from adjacent locations of mining area water quality, mineralogical and geochemical studies. Mularz mapped the problem of environmental monitoring and land-use/land cover changes over the lignite open-cast mine and power plant area was investigated using airborne remote photography along with Landsat TM and SPOT imageries in the central part of Poland to discriminate, assess and even to measure these destructive phenomena.

The degradation of land use due to coal mining using remote sensing techniques at Jharia coal filed have been studied by Prakash and Gupta. The open cast mining activities like lignite and other materials lead to loss of fertile agricultural land, elimination of surface water bodies and ground water depletion in deeper aquifers. Oxidation process at surface of dumped mine waste may produce acid water drainage, which can affect the surface and groundwater quality.

Scientists have successfully utilized the hyperspectral sensor data to map the mineral abundance, lithological mapping and processing methodology for detecting iron and manganese mines in parts of Singhbum district, Orissa and spectral signature and spectral mixture modelling techniques utilized for targeting laterite & bauxite ore deposits. Levesque et al., used the hyperspectral remote sensing data to monitor and assess the rehabilitation of mine tailing sites. Chevrel et al, effectively utilized airborne hyperspectral remote sensing sensors in the six mining areas (Europe and Greenland) to study the mining related contamination and its impact on vegetation. Different primary granites and subsequent kaolinization in the mining area were identified using hyperspectral data analysis. Lalan kumar et al, applied the application of Geographical Information system to underground mining studies including land ownership and mineral claims, exploration management production and mine site.

Multi-date infra red Landsat images were utilized to study the environmental changes in Sierra Leone, West Africa, especially to understand the impact on hydrogeomorphology.

An attempt has made to delineate the magnesite ore deposits in Salem using hyperspectral remote sensing data, which reveals that potential of using narrow band hyperspectral data for further mapping of impact mining on environment. The management and controlling factors of environment affected due to mining to be adopted both during production and after closure. Sufficient data collection and accurate processing should be done with respect to place and time for control and planning the environmental management.

Urban Environment Management

Urbanization is an index of transformation from traditional rural economies to modern industrial one. It is a progressive concentration of population in urban unit. During the last fifty years the global population is increased dramatically, as a result most urban settlements are characterized by shortfalls in stock housing and water supply, urban encroachments in fringe area, inadequate sewerage, traffic congestion, pollution, poverty and social unrest making urban governance a difficult task to maintain healthy urban environment. High rate of urban population growth is a cause of concern among urban and town planners for efficient urban planning. Therefore, there is an urgent need to adopt modern technology of remote sensing which includes both aerial as well as satellite based systems, allowing us to collect lot of physical data rather easily, with speed and on repetitive basis, and together with GIS helps us to analyze the data spatially.

Floods cause damage to natural resources and environmental quality and indirectly contribute to increasing poverty, which in turn further add to the vulnerability of both natural and human systems mostly urban area compare to the rural areas. The environment and flood linkage has been recognized, and many environmental programs such as reforestation, forest protection, upland fixed cultivation and resettlement, could be been implemented through remote sensing and GIS.

GIS has been widely used in characterization and assessment studies which require a watershed-based approach to manage the water level and waste management in the urban locations. Basic physical characteristics of a watershed such as the drainage network and flow paths can be derived from readily available Digital Elevation Models (DEMs). When faced with challenges involving water quality and quantity due to natural as well as human-induced hazards (e.g., droughts, hazard material spills, floods, and urbanization), planning becomes extremely important so as to mitigate their impacts and ensure optimal utilization of the available resources.

Remote sensing can provide an important source of data for urban land use/land cover mapping and environmental monitoring. A numbers of significant studies were made for environmental quality management. Uncontrolled urbanization has been responsible for several problems, our cities facing today, resulting in substandard living environment, acute problems of drinking water, noise and air pollution, disposal of waste, traffic congestion etc. To minimise these environmental degradations in and around cities, the technological development in related fields have to address to these problems caused by rapid urbanization, only then the fruits of development will percolate to the most deprived ones. The modern technology of remote sensing which includes both aerial as well as satellite based systems, allow us to collect physical data rather easily, with speed and on repetitive basis, and together with GIS helps us to analyze the data spatially, offering possibilities of generating various options (modeling), thereby optimizing the whole planning process.

The dynamic nature of urban environmental necessitates both macro and micro level analysis. Therefore, it is necessary for policy makers to integrate remote sensing with urban planning and management. The trend towards using remotely sensed data in urban studies began with first-generation satellite sensors such as Landsat MSS and was given impetus by a number of second generation Satellites: Landsat TM, ETM+ and SPOT. The recent advent of a third generation of very high spatial resolution (5m/pixel) satellite sensors is stimulating. The high resolution PAN and LISS III merged data may be used together effectively for urban applications. Data from IRS P-6 satellites with sensors on board especially LISS IV Mono and Multispectral (MX) with 5.8 m/pixel spatial resolution is very useful for intensive urban studies.

Coastal and Marine Environment

Coastal zones in are constantly undergoing wide-ranging changes in shape and environment due to natural as well as human development activities. Natural processes such as waves, erosion, changes in river courses etc., cause long time effect at slower rate; but manmade activities, such as settlement, industrial activities, recreational activities, waste disposal etc., affect the coastal environment at comparatively much faster rate. Continued loss of these wetlands may lead to the collapse of coastal ecosystems. It is, therefore, necessary to monitor coastal zone changes with time. Remote sensing technology in recent years has proved to be of great importance in acquiring data for effective resources management and hence could also be applied to coastal environment monitoring and management. The high temporal resolution provided by the satellite data is found to be a major improvement in studying the behavior of suspended sediments in the coastal waters, which would help in understanding the movement of sediments and pollutants.

GIS in addition to providing efficient data storage and retrieval facilities also offers a cheaper option of monitoring forest conditions over time. Remote sensing and GIS are increasingly used in mangrove forestry worldwide to assist in gathering and analysing images acquired from aircrafts, satellites and even balloons. The notable advantages of using GIS include the ability to update the information rapidly, to undertake comparative analytical work and making this information available as required. The area covered by mangroves in the islands of Andaman was calculated using SPOT 1993 and

IRS 1D LISS III 2003 imageries. Twumasi and Merem assessed change within a coastal environment in the Niger delta region of Nigeria using remotely sensed satellite imagery and GIS modeling, quickened the analysis of the spatial distribution of environmental change involving land use, land cover classification, forest and hydrology and demographic issues facing the Niger Delta and successful implemented some of the strategies could lead to effective management of the coastal environment in the Niger Delta region.

Satellite based remote sensing techniques have proved successful in providing a comprehensive, reliable and up-to date information on land use/land cover in the offshore areas of east coast of Andhra Pradesh in the most cost effective manner. Environmental Sensitivity Index (ESI) and Reach Sensitivity Index (RSI) identified through modern methods like Digital Image processing and GIS for preparedness in case of oil spill incidents in offshore areas. Satellite based remote sensing techniques have proved successful in providing a comprehensive, reliable and up-to date information on land use/land cover in the offshore areas of east coast of Andhra Pradesh in the most cost effective manner. Environmental Sensitivity Index (ESI) and Reach Sensitivity Index (RSI) identified through modern methods like Digital Image processing and GIS for preparedness in case of oil spill incidents in offshore areas.

The combination of remote sensing and GIS technologies provides an ideal solution for understanding the spatial/temporal distribution of oil spills in the marine environment and is considered as the core of the oil spill monitoring system. The advantages of the remote sensing and GIS provides the ability to extract the oil pollution parameters such as location and spill areas including spatial and temporal information allows the users to establish the major cause and source of oil spills and then outline the risk areas to save the marine environment. One of the major advantages of GIS is the ability to extract oil pollution parameters such as location, size and spill areas. Spatial and temporal information (oil spill distribution at sea and its evolution in time) allows the users to establish the major cause and source of oil spills, and then outline the risk area.

The products derived from geospatial technologies support informed decision making with respect to marine spatial planning and management.

Wasteland Environment

Wetlands consist of 3 - 6% of the earth's land surface, while they make available supplies and services such as: water quality maintenance, agricultural production, fisheries, and recreation floodwater, retention, provision of wildlife habitat, and control of soil erosion. Wetlands are transitional lands between Terrestrial and aquatic system that provide many goods and services including flood water retention, water quality maintenance, wildlife habitat, and soil erosion control. To prevent further loss of wetlands, and conserve existing wetland ecosystem for biodiversity and ecosystem services and goods, it is important to inventory and monitor wetlands and their adjacent uplands. The area of wetlands are reducing constantly for the last few decades due to wetland reclamation, population pressure, water diversion, dam construction, pollution, biological incursion, desertification, climate change, and misguiding policies. Remotely sensed data have been utilized to measure the qualitative and quantitative terrestrial land-cover changes=. During last two decades a diversity of remotely sensed data and change detection methods have been developed and assessed. Remote sensing (RS) data and Geographic information systems (GIS) are appropriate tools for monitoring of the wetland distribution area and spatial-temporal dynamic multiplicity Satellite remote sensed data have been widely utilized for inventorying and monitoring wetlands and can also provide information on surrounding land use and their change over the time successfully utilized the Multi-temporal remote sensing data and GIS for wetland mapping in the southwest of Iran near to the Karkheh River using four Landsat images 1985 (Landsat MSS), 1999 (Landsat ETM+), 2002 (Landsat ETM+) and 2011 (Landsat ETM+) and found that, increase in agricultural activity, climate change and construction engineering projects caused wetland surface area reduction.

Satellite remote sensing has many advantages for inventory and monitoring of wetlands and also provide information on surrounding land use and their changes over time. Landsat MSS, TM, and SPOT are common data type for wetland classification and its spatial-temporal dynamic change.

Due to temporal revisit capability of the satellite data, it allows to monitor the wetlands either seasonally or yearly. The use of remote sensing data for land cover classification is less costly and less time-consuming than aerial photography for large geographic areas. For wetland studies such as monitoring and inventory use and apply satellite remote sensed data can suitable in developing countries, where the budget are restricted and the data about the wetland like wetland area, land use, and wetland losses are limited.

Remote sensing has served as an efficient method of gathering data about glaciers since its emergence. The recent advent of Geographic Information Systems (GIS) and Global Positioning Systems (GPS) has created an effective means by which the acquired data are analyzed for the effective monitoring and mapping of temporal dynamics of glaciers. A large number of researchers have taken advantage of remote sensing, GIS and GPS in their studies of glaciers.

Study of Natural Hazards

Disasters can cause drastic environmental changes. A large amount of spatial data is required for managing the disasters and to assess their environmental impacts. Earth observation data offers independent coverage of wide areas for a broad spectrum of crisis situations. It provides information over large areas in near-real-time interval and supplementary at short-time and long-time intervals. Therefore, remote sensing can support disaster management in various applications.

The impact of disasters on the environment has become more severe over the last decades. Moreover, the reported number of disasters has dramatically increased, as well as the costs to the global economy and the number of people affected. The reasons for these disasters are manifold, and the impact can be found in the increasing vulnerability of societies, infrastructure, and population. Furthermore, extreme weather events have become more common and severe.

The increasing occurrences of natural and man-made disasters lead to a growing demand for up-to-date geographic information, especially timely material on rapidly evolving events. This includes comprehensive, near-real-time Earth observation data, which offer independent coverage of wide areas for a broad spectrum of civilian crisis situations. Satellite imagery can serve as a source of information in disaster situation. Accordingly, remote sensing can provide information on various domains of the disaster management, from risk modelling and vulnerability analysis to early warning and damage assessment.

Disaster Management and Remote Sensing

Disaster Types and their Environmental Impact

There are several ways to classify disaster types. One common classification is natural and man-made disasters. Severe geo-physical or climatic events, such as volcanic eruptions, floods, cyclones and fires that threaten people or property, are termed as natural disasters. Man-made disasters are events which are caused by human activities (e.g. industrial chemical accidents and oil spills). Sometimes, natural disasters that are accelerated by human influence are termed human-induced disasters. In addition, the Centre for Research on the Epidemiology of Disasters divides the natural disaster category into six sub-groups, which in turn include 17 disaster types, and 33 sub-types. The technological disaster category is segregated into three sub-groups which in turn include 15 disaster types. Besides, disasters can be categorised as acute (e.g. earthquake) or slow (e.g. drought) based on their onset.

Table: Natural disasters categorisation.

Natural Disaster Sub-group			
Climatological	Geophysical	Hydrological	Meterological
Natural Disaster Types and Sub-types			
Drought	Earthquake.	Flood.	Storm.
Glacial Lake Outburst	Ground Shaking.	Coastal food.	Extra-tropical cyclone.

Wildfire	Tsunami.	Riverine flood.	Tropical cyclone.
Forest fire	Mass movement.	Flash flood.	Convective Storm.
Land fire	Volcanic activity.	Ice jam flood.	Extreme temperature.
	Ash fall.	Landslide.	Cold wave.
	Lahar.	Avalanche(snow, debris, mudflow, rockfall).	Heat wave.
	Pyroclastic flow.	Wave action.	Severe winter condition.
	Lawa flow.	Rogue wave.	Fog.
		Seiche.	

Table: Man-made disasters categorisation.

Man-made Disaster Sub-group		
Industrial accident	Transport accident	Miscellaneous accident
Man-made Disaster Types		
Chemical spill	Air	Collapse
Collapse	Road	Explosion
Explosion	Rail	Fire
Fire	Water	Other
Gas leak		
Poisoning		
Radiation		
Othero		

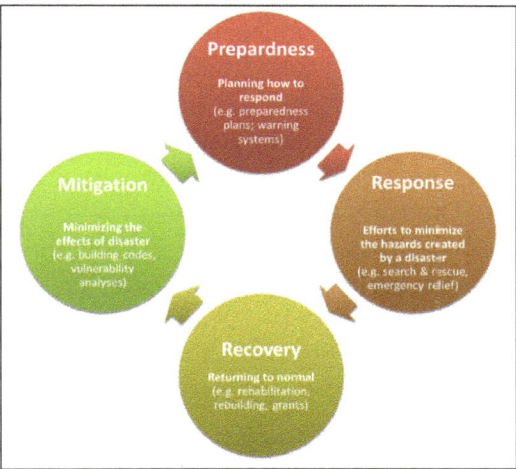

Disaster management cycle.

There are many effects that result from disasters, whether natural or man-made. For instance, the impacts of disasters have a human and an environmental dimension. UNEP concludes that 'environmental conditions may exasperate the impact of a disaster, and vice versa, disasters tend to have an impact on the environment. They raise the interesting point that while most environmental impacts are negative, some are positive. For example, 'floods can help rejuvenate floodplain vegetation and are important drivers of many ecological processes in floodplains'.

The impact of disasters can be reduced through a proper disaster management. The process of disaster management is often interpreted as a cycle consisting of four main phases: mitigation, preparedness, response and recovery.

Earth Observation for Disaster Management: Potential and Limitations

The Earth is being imaged each day by a constellation of remote sensing satellites. Two complementary types of Earth observation satellites are particularly relevant to disaster management. 'Geostationary Earth observation satellites' are placed at an altitude of approximately 35,800 kilometres. At this altitude, one orbit takes 24 hours, the same length of time as the Earth requires to rotate once on its axis. In effect, it means that satellites in this orbit remain stationary above the ground and view the whole Earth disk below. Their spatial data resolution is very low but is collected at the same point every 15 minutes. With these kinds of data, the evolution of atmospheric phenomena can be observed, ensuring real-time coverage of meteorological events such as severe local storms and tropical cyclones. The importance of this capability has been exemplified during several hurricane events.

The great advantage of 'polar-orbiting satellites' is the provision of relatively high spatial resolution data (up to 0.3 meter for optical imagery and 1 meter for radar imagery), which is very important for mapping disaster damages in detail, such as affected infrastructure or buildings after an earthquake. Most of the Earth observation satellites are in a low and 'near-polar' orbit with an orbital period of approximately 90–100 minutes and an orbit inclination near 90 degrees. This allows the satellite to see virtually every part of the Earth as the Earth rotates underneath it. However, no spot on the Earth's surface can be sensed continuously or at any point of time from a satellite in a polar orbit. The time elapsed between observations of the same point on the Earth (revisit time) is limited to once every few days with the same sensor parameters or maximum once a day for steerable satellite. Moreover, most satellites do not continuously collect data due to limitations in power and memory. Some offer regular and reliable data acquisition while others may be more ad hoc, collecting only 5 or 10 minutes' worth of data in a 90-minute orbit. Data are stored on board the satellite until it is in sight of a ground station to downlink the data. The time between an image being taken and being available to download can range between a month to a few minutes and is getting faster all the time. Thus, the collection of high-resolution data has some limitations regarding acquisition time, data provision and image extent.

The Earth observation satellites have their own special systems of imaging sensors which make use of the visible, infrared, microwave and other parts of the electromagnetic spectrum.

Optical data are of great importance for disaster management support, because they can be used nearly for all disaster types and for all phases of disaster management. For example, they are used for planning the logistics of relief actions in the field immediately after an earthquake or tsunami. Optical images are easy to understand and interpret even for non-specialists, particularly when it consists of the three visual primary colour bands (red, green and blue) and the bands are combined to produce a 'true colour' image. However, the interpretation of false colour composite images is not intuitive and requires expert knowledge; likewise, all advanced analysis techniques need comprehensive know-how. To select the most appropriate data type for the needs of the individual disaster situation, the characteristics of the sensor are of great importance. Particularly, temporal and spatial resolutions are key factors. For example, for mapping an earthquake in an urban area optical data with a spatial resolution of <0.5 meters are most valuable. The most crucial point for the use of optical images is their availability. Due to cloud coverage, haze and other atmospheric conditions useful optical images could not obtained by every satellite overpass. Aggravating this situation, there are some disasters such as wildfire or severe storms which are characterised by clouds and smoke.

The thermal imagery offers excellent possibilities for automated extraction of anomalous high temperature or hot spots caused by wild fires or information about volcanic eruptions. However, due to the fact that energy decreases with increasing wavelength, thermal wavelength have relatively low energy levels and consequently thermal image data have a lower spatial resolution than optical data. Techniques for automatic fire detection from the space are operational and are accepted by the users (e.g. European Forest Fire Information System).

Microwave sensors are of great value for the fast response mapping and analysis tasks, as they allow imaging at wavelengths almost unaffected by atmospheric disturbances such as rain or cloud. Most modern synthetic aperture radar (SAR) sensors are designed to acquire data from various ground resolution elements. In most applications, only the relative variability of backscatter intensity within the image is used. Nonetheless, backscatter intensity and the phase of SAR images can be utilised. Phase information of a single SAR data set has no value, but the comparison of phases between two SAR images acquired at distinct times are utilised in SAR interferometry or INSAR. Moreover, with modern satellites (e.g. TerraSAR-X and Radarsat-2) it is possible to acquire simultaneous data with more than one polarisation. SAR systems can be used to map flooding or to measure earth deformations before and during earthquakes or volcanic eruptions, particularly when post-event imagery can be jointly analysed with archived reference imagery for change detection or interferometric coherence or displacement measurements.

In general, the availability of appropriate data with respect to acquisition time, image extent, spatial as well as temporal and spectral resolution is an important consideration for most applications in the disaster context. Particularly, there are numerous examples for the importance of the necessity of fast availability of remote sensing data like damage assessment maps for earthquakes, landslides or flooding. However, for monitoring the spread of an oil spill or the extent of flooding the revisit time is relevant too.

Remote sensing has proven to be useful for a range of applications. Especially high spatial resolution data and remote sensing techniques are being deployed in the context of the disaster management domain, from risk modelling and vulnerability analysis to early warning and damage assessment.

Rapid Mapping Work Flow

No decision maker or relief worker can work with raw satellite imagery. To generate the required situation maps, reports or statistics, which can be read and understood by non-satellite expert users, experts in remote sensing and cartography are necessary. In 2004, German Aerospace Center (DLR) was one of the first institutions, which has set up a dedicated interface called Center for Satellite Based Crisis Information (ZKI) to facilitate the use of its Earth observation capacities in the services of national and international response to disaster situations. ZKI's function is particularly 'rapid mapping' – the rapid acquisition, processing and analysis of satellite data and the provision of satellite-based information products. Analyses are tailored to meet the specific requirements of national and international political bodies or humanitarian relief organisations. In order to provide up-to-date and relevant satellite-based cartographic information and situation analysis, it is necessary to establish efficient and operational data flow lines between satellite operators, receiving stations and distribution networks on the one hand and the decision makers and relief workers on the other hand. Service lines and feedback loops have been created to allow best possible data and information provision, as well as optimised decision support. In order to meet with users' demands and service requirements in crisis situations, ZKI set up a rapid mapping workflow ensuring a fast access to available, reliable and affordable crisis information worldwide.

Schedules for the full cycle from the emergency call (mobilisation phase), satellite tasking (data acquisition), pre-processing, analysis and interpretation, map production and data provision to the end-user are tight (as fast as possible). Hence, rapid mapping is still a complex task.

After the mandatory decision process, whether satellite analysis is appropriate for the respective crisis or not, the area of interest has to be defined and cross-checked to avoid false geolocation. Following this iterative process, it has to be assured that all applicable satellites are programmed for data acquisition. Furthermore, an enquiry for corresponding archive imagery has to be set up for documentation of the pre-disaster situation and change detection analysis. Besides the procurement of satellite data, it is necessary to check and prepare supplementary geodata such as population and infrastructure data, road network, contour lines and administrative boundaries. Experience of several activations and user feedback shows that additional geoinformation increases the satellite data analysis significantly. This includes place names, critical infrastructure, transportation network or further detailed specifications. Availability and access to accurate and up-to-date spatial data, particularly in remote regions, are the most crucial problems.

After receiving the archived and recently recorded satellite imagery, essential pre-processing has to be done. This includes geo- and ortho-rectification as well as radiometric corrections and data format conversions. Data re-projection is necessary due to varying demands and standards. In the majority of activations, a Universal Transverse Mercator (UTM) projection is used due to global applicability and following international standards. Depending on user's needs, crisis type and extent, different analysis process chains have to be applied.

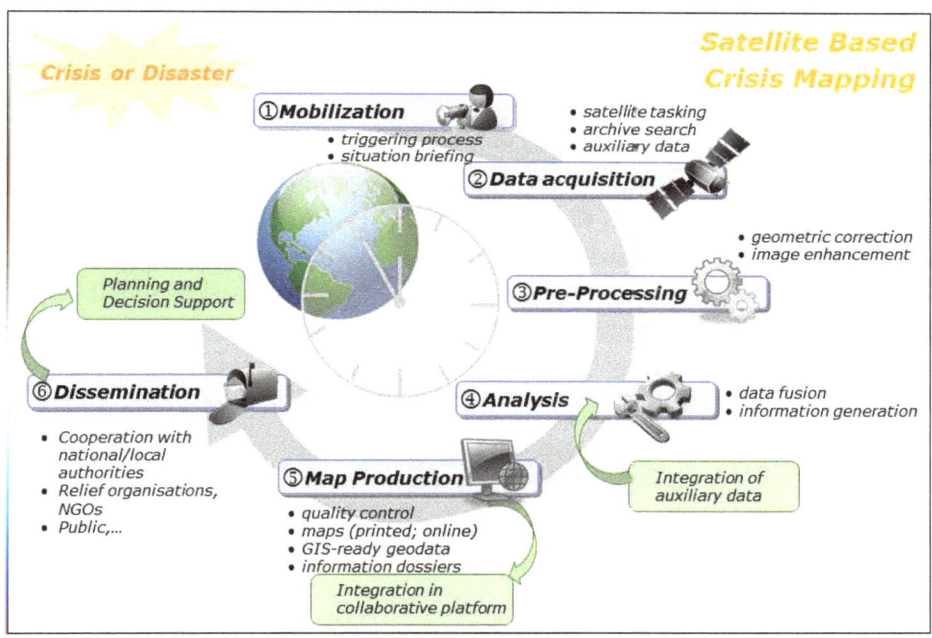

Rapid mapping workflow.

Derivation of water surfaces or general damage assessment is dependent on input data type, scale and possible availability of archived satellite imagery. Before and after image comparison allows the quantification of affected areas. This change detection method can either be applied for optical or radar imagery in order to detect areas where significant change can be identified. Furthermore, general image classification and differencing methods allows quantification of flooded areas, fire scars or damaged areas.

Situation and damage maps are generated in order to translate complex satellite information in readable and coherent crisis information. Following this map compilation, an adapted map generation process is applied. A settled quality control process takes place after each single product generation step as well as before publishing. Delivery is accomplished via Internet, intranet, ftp, e-mail or satellite communication. Furthermore, printed and laminated maps will be sent via express delivery on request. User feedback from field units has proved to be an important source for optimisation. Maps are updated when new and improved data are available or knowledgeable feedback is received even though the maps are published and delivered.

In order to fulfil its tasks, DLR-ZKI is involved in international, European and national mechanisms providing space-based information supporting the disaster relief (e.g. International Charter Space and Major Disaster). The understanding of the organisational frameworks of these mechanisms, their activation procedures and workflows are a prerequisite to take advantage of the products provided by these mechanisms.

Hyperspectral Imaging

Hyperspectral imaging, like other spectral imaging, collects and processes information from across the electromagnetic spectrum. The goal of hyperspectral imaging is to obtain the spectrum for each pixel in the image of a scene, with the purpose of finding objects, identifying materials, or detecting processes. There are three general branches of spectral imagers. There are push broom scanners and the related whisk broom scanners (spatial scanning), which read images over time, band sequential scanners (spectral scanning), which acquire images of an area at different wavelengths, and snapshot hyperspectral imaging, which uses a staring array to generate an image in an instant.

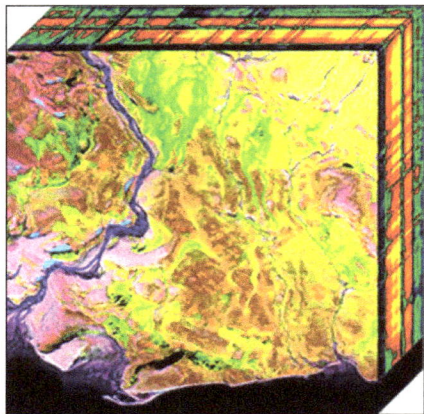

Two-dimensional projection of a hyperspectral cube.

Whereas the human eye sees color of visible light in mostly three bands (long wavelengths - perceived as red, medium wavelengths - perceived as green, and short wavelengths - perceived as blue), spectral imaging divides the spectrum into many more bands. This technique of dividing images into bands can be extended beyond the visible. In hyperspectral imaging, the recorded spectra have fine wavelength resolution and cover a wide range of wavelengths. Hyperspectral imaging measures continuous spectral bands, as opposed to multispectral imaging which measures spaced spectral bands.

Engineers build hyperspectral sensors and processing systems for applications in astronomy, agriculture, molecular biology, biomedical imaging, geosciences, physics, and surveillance. Hyperspectral sensors look at objects using a vast portion of the electromagnetic spectrum. Certain objects leave unique 'fingerprints' in the electromagnetic spectrum. Known as spectral signatures, these 'fingerprints' enable identification of the materials that make up a scanned object. For example, a spectral signature for oil helps geologists find new oil fields.

Hyperspectral Image Sensors

Figuratively speaking, hyperspectral sensors collect information as a set of 'images'. Each image represents a narrow wavelength range of the electromagnetic spectrum, also known as a spectral band. These 'images' are combined to form a three-dimensional (x,y,λ) hyperspectral data cube for processing and analysis, where x and y represent two spatial dimensions of the scene, and λ represents the spectral dimension (comprising a range of wavelengths).

Technically speaking, there are four ways for sensors to sample the hyperspectral cube: Spatial scanning, spectral scanning, snapshot imaging, and spatio-spectral scanning.

Hyperspectral cubes are generated from airborne sensors like the NASA's *Airborne Visible/Infrared Imaging Spectrometer* (AVIRIS), or from satellites like NASA's EO-1 with its hyperspectral instrument Hyperion However, for many development and validation studies, handheld sensors are used.

The precision of these sensors is typically measured in spectral resolution, which is the width of each band of the spectrum that is captured. If the scanner detects a large number of fairly narrow frequency bands, it is possible to identify objects even if they are only captured in a handful of pixels. However, spatial resolution is a factor in addition to spectral resolution. If the pixels are too large, then multiple objects are captured in the same pixel and become difficult to identify. If the pixels are too small, then the energy captured by each sensor cell is low, and the decreased signal-to-noise ratio reduces the reliability of measured features.

The acquisition and processing of hyperspectral images is also referred to as imaging spectroscopy or, with reference to the hyperspectral cube, as 3D spectroscopy.

Technologies for Hyperspectral Data Acquisition

There are *four basic techniques* for acquiring the three-dimensional (x,y,λ) dataset of a hyperspectral cube. The choice of technique depends on the specific application, seeing that each technique has context-dependent advantages and disadvantages.

Individual sensor outputs for the four hyperspectral imaging techniques.
From left to right: Slit spectrum; monochromatic spatial map; 'perspective projection' of hyperspectral cube; wavelength-coded spatial map.

Spatial Scanning

In spatial scanning, each two-dimensional (2-D) sensor output represents a full slit spectrum (x,λ). Hyperspectral imaging (HSI) devices for spatial scanning obtain slit spectra by projecting a strip of the scene onto a slit and dispersing the slit image with a prism or a grating. These systems have the drawback of having the image analyzed per lines (with a push broom scanner) and also having some mechanical parts integrated into the optical train. With these *line-scan systems*, the spatial dimension is collected through platform movement or scanning. This requires stabilized mounts or accurate pointing information to 'reconstruct' the image. Nonetheless, line-scan systems are particularly common in remote sensing, where it is sensible to use mobile platforms. Line-scan systems are also used to scan materials moving by on a conveyor belt. A special case of line scanning is *point scanning* (with a whisk broom scanner), where a point-like aperture is used instead of a slit, and the sensor is essentially one-dimensional instead of 2-D.

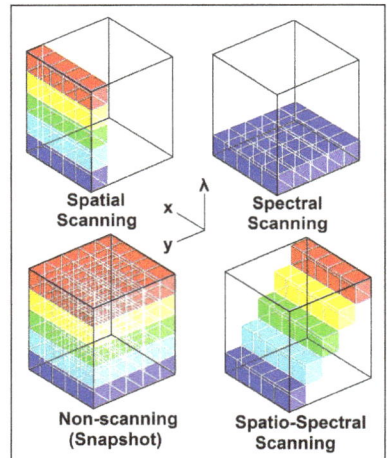

Acquisition techniques for hyperspectral imaging, visualized as sections of the hyperspectral datacube with its two spatial dimensions (x,y) and one spectral dimension (lambda).

Spectral Scanning

In spectral scanning, each 2-D sensor output represents a monochromatic ('single-colored'), spatial (x,y) map of the scene. HSI devices for spectral scanning are typically based on optical band-pass filters (either tunable or fixed). The scene is spectrally scanned by exchanging one filter after another while the platform remains stationary. In such 'staring', wavelength scanning systems, spectral smearing can occur if there is movement within the scene, invalidating spectral correlation/detection. Nonetheless, there is the advantage of being able to pick and choose spectral bands, and having a direct representation of the two spatial dimensions of the scene. If the imaging system is used on a moving platform, such as an airplane, acquired images at different wavelengths corresponds to different areas of the scene. The spatial features on each of the images may be used to realign the pixels.

Non-scanning

In non-scanning, a single 2-D sensor output contains all spatial (x,y) and spectral (λ) data. HSI devices for non-scanning yield the full datacube at once, without any scanning. Figuratively speaking, a single snapshot represents a perspective projection of the datacube, from which its three-dimensional structure can be reconstructed. The most prominent benefits of these snapshot hyperspectral imaging systems are the *snapshot advantage* (higher light throughput) and shorter acquisition time. A number of systems have been designed, including computed tomographic imaging spectrometry (CTIS), fiber-reformatting imaging spectrometry (FRIS), integral field spectroscopy with lenslet arrays (IFS-L), multi-aperture integral field spectrometer (Hyperpixel Array), integral field spectroscopy with image slicing mirrors (IFS-S), image-replicating imaging spectrometry (IRIS), filter stack spectral decomposition (FSSD), coded aperture snapshot spectral imaging (CASSI), image mapping spectrometry (IMS), and multispectral Sagnac interferometry (MSI). However, computational effort and manufacturing costs are high. In an effort to reduce the computational demands and potentially the high cost of non-scanning hyperspectral instrumentation, prototype devices based on Multivariate Optical Computing have been demonstrated. These devices have been based on the Multivariate Optical Element spectral calculation engine or the Spatial Light Modulator spectral calculation engine. In these platforms, chemical information is calculated in the optical domain prior to imaging such that the a chemical image relies on conventional camera systems with no further computing. As a disadvantage of these systems, no spectral information is ever acquired, i.e. only the chemical information, such that post processing or reanalysis is not possible.

Spatiospectral Scanning

In spatiospectral scanning, each 2-D sensor output represents a wavelength-coded ('rainbow-colored', $\lambda=\lambda(y)$), spatial (x,y) map of the scene. A prototype for this technique, introduced in 2014, consists of a camera at some *non-zero* distance behind a basic slit spectroscope (slit + dispersive element). Advanced spatiospectral scanning systems can be obtained by placing a dispersive element before a spatial scanning system. Scanning can be achieved by moving the whole system relative to the scene, by moving the camera alone, or by moving the slit alone. Spatiospectral scanning unites some advantages of spatial and spectral scanning, thereby alleviating some of their disadvantages

Distinguishing Hyperspectral from Multispectral Imaging

Hyperspectral imaging is part of a class of techniques commonly referred to as spectral imaging or spectral analysis. Hyperspectral imaging is related to multispectral imaging. The distinction between hyper- and multi-spectral is sometimes based incorrectly on an arbitrary "number of bands" or on the type of measurement. Hyperspectral imaging (HSI) uses continuous and contiguous ranges of wavelengths (e.g. 400 - 1100 nm in steps of 0.1 nm) whilst multispectral imaging (MSI) uses a subset of targeted wavelengths at chosen locations (e.g. 400 - 1100 nm in steps of 20 nm).

Multispectral imaging deals with several images at discrete and somewhat narrow bands. Being "discrete and somewhat narrow" is what distinguishes multispectral imaging in the visible wavelength from color photography. A multispectral sensor may have many bands covering the spectrum from the visible to the longwave infrared. Multispectral images do not produce the "spectrum" of an object. Landsat is an excellent example of multispectral imaging.

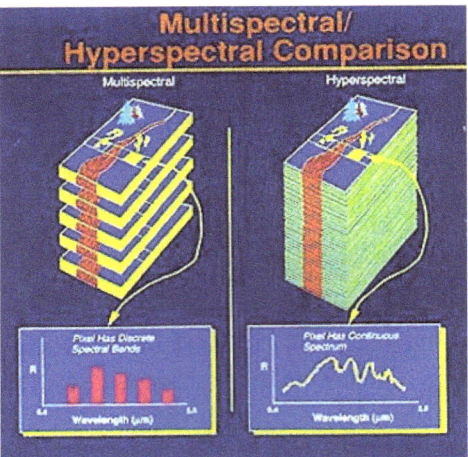

Hyperspectral and Multispectral Differences.

Hyperspectral deals with imaging narrow spectral bands over a continuous spectral range, producing the spectra of all pixels in the scene. A sensor with only 20 bands can also be hyperspectral when it covers the range from 500 to 700 nm with 20 bands each 10 nm wide. (While a sensor with 20 discrete bands covering the VIS, NIR, SWIR, MWIR, and LWIR would be considered multispectral.).

'Ultraspectral' could be reserved for interferometer type imaging sensors with a very fine spectral resolution. These sensors often have (but not necessarily) a low spatial resolution of several pixels only, a restriction imposed by the high data rate.

Applications

Hyperspectral remote sensing is used in a wide array of applications. Although originally developed for mining and geology (the ability of hyperspectral imaging to identify various minerals makes it ideal for the mining and oil industries, where it can be used to look for ore and oil), it has now spread into fields as widespread as ecology and surveillance, as well as historical manuscript research, such as the imaging of the Archimedes Palimpsest. This technology is continually becoming more available to the public. Organizations such as NASA and the USGS have catalogues of various minerals and their spectral signatures, and have posted them online to make them readily available for researchers. On a smaller scale, NIR hyperspectral imaging can be used to rapidly monitor the application of pesticides to individual seeds for quality control of the optimum dose and homogeneous coverage.

Agriculture

Although the cost of acquiring hyperspectral images is typically high, for specific crops and in specific climates, hyperspectral remote sensing use is increasing for monitoring the development and health of crops. In Australia, work is under way to use imaging spectrometers to detect grape variety and develop an early warning system for disease outbreaks. Furthermore, work is underway to use hyperspectral data to detect the chemical composition of plants, which can be used to detect the nutrient and water status of wheat in irrigated systems. On a smaller scale, NIR hyperspectral imaging can be used to rapidly monitor the application of pesticides to individual seeds for quality control of the optimum dose and homogeneous coverage.

Hyperspectral camera embedded on OnyxStar HYDRA-12 UAV from AltiGator.

Another application in agriculture is the detection of animal proteins in compound feeds to avoid bovine spongiform encephalopathy (BSE), also known as mad-cow disease. Different studies have been done to propose alternative tools to the reference method of detection, (classical microscopy). One of the first alternatives is near infrared microscopy (NIR), which combines the advantages of microscopy and NIR. In

2004, the first study relating this problem with hyperspectral imaging was published. Hyperspectral libraries that are representative of the diversity of ingredients usually present in the preparation of compound feeds were constructed. These libraries can be used together with chemometric tools to investigate the limit of detection, specificity and reproducibility of the NIR hyperspectral imaging method for the detection and quantification of animal ingredients in feed.

Eye Care

Researchers at the Université de Montréal are working with Photon etc. and Optina Diagnostics to test the use of hyperspectral photography in the diagnosis of retinopathy and macular edema before damage to the eye occurs. The metabolic hyperspectral camera will detect a drop in oxygen consumption in the retina, which indicates potential disease. An ophthalmologist will then be able to treat the retina with injections to prevent any potential damage.

Food Processing

In the food processing industry, hyperspectral imaging, combined with intelligent software, enables digital sorters (also called optical sorters) to identify and remove defects and foreign material (FM) that are invisible to traditional camera and laser sorters. By improving the accuracy of defect and FM removal, the food processor's objective is to enhance product quality and increase yields.

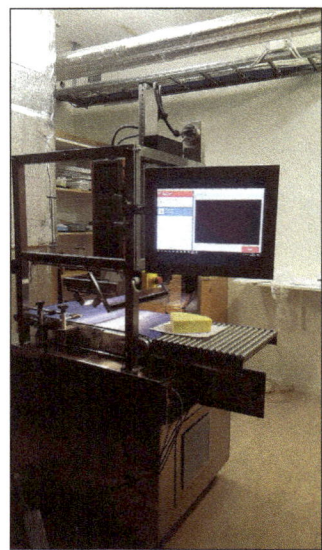

A line scan push-broom system was used to scan the cheeses and images were acquired using a Hg-Cd-Te array (386x288) equipped linescan camera with halogen light as a radiation source.

Adopting hyperspectral imaging on digital sorters achieves non-destructive, 100 percent inspection in-line at full production volumes. The sorter's software compares the hyperspectral images collected to user-defined accept/reject thresholds, and the ejection system automatically removes defects and foreign material.

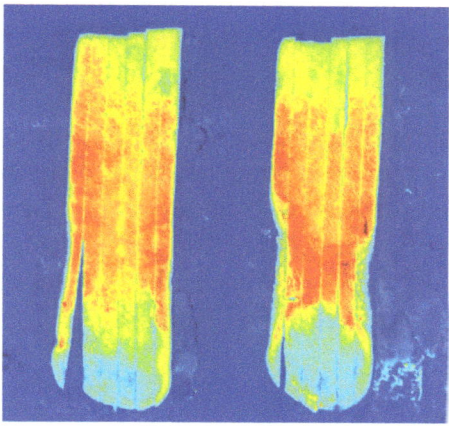

Hyperspectral image of "sugar end" potato strips shows invisible defects.

The recent commercial adoption of hyperspectral sensor-based food sorters is most advanced in the nut industry where installed systems maximize the removal of stones, shells and other foreign material (FM) and extraneous vegetable matter (EVM) from walnuts, pecans, almonds, pistachios, peanuts and other nuts. Here, improved product quality, low false reject rates and the ability to handle high incoming defect loads often justify the cost of the technology.

Commercial adoption of hyperspectral sorters is also advancing at a fast pace in the potato processing industry where the technology promises to solve a number of outstanding product quality problems. Work is underway to use hyperspectral imaging to detect "sugar ends," "hollow heart" and "common scab," conditions that plague potato processors.

Mineralogy

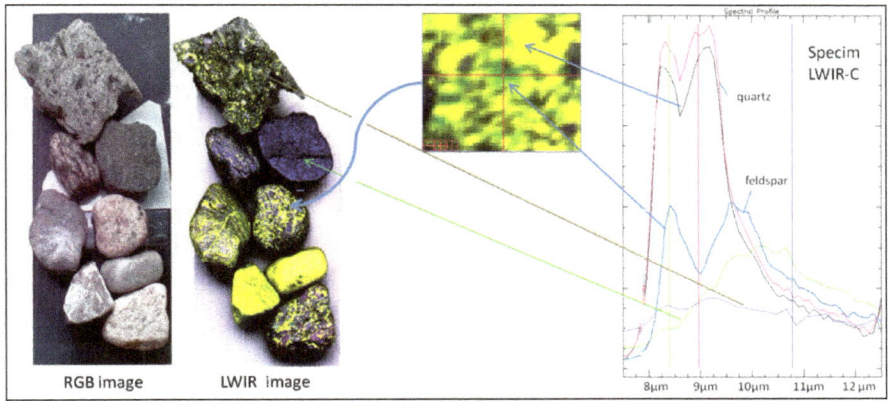

A set of stones is scanned with a Specim LWIR-C imager in the thermal infrared range from 7.7 μm to 12.4 μm. The quartz and feldspar spectra are clearly recognizable.

Geological samples, such as drill cores, can be rapidly mapped for nearly all minerals of commercial interest with hyperspectral imaging. Fusion of SWIR and LWIR spectral imaging is standard for the detection of minerals in the feldspar, silica, calcite, garnet, and olivine groups, as these minerals have their most distinctive and strongest spectral signature in the LWIR regions.

Hyperspectral remote sensing of minerals is well developed. Many minerals can be identified from airborne images, and their relation to the presence of valuable minerals, such as gold and diamonds, is well understood. Currently, progress is towards understanding the relationship between oil and gas leakages from pipelines and natural wells, and their effects on the vegetation and the spectral signatures. Recent work includes the PhD dissertations of Werff and Noomen.

Surveillance

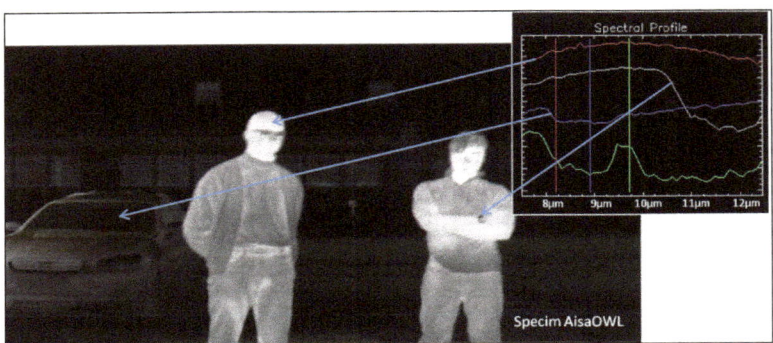

Hyperspectral thermal infrared emission measurement, an outdoor scan in winter conditions, ambient temperature -15 °C—relative radiance spectra from various targets in the image are shown with arrows. The infrared spectra of the different objects such as the watch glass have clearly distinctive characteristics. The contrast level indicates the temperature of the object. This image was produced with a Specim LWIR hyperspectral imager.

Hyperspectral surveillance is the implementation of hyperspectral scanning technology for surveillance purposes. Hyperspectral imaging is particularly useful in military surveillance because of countermeasures that military entities now take to avoid airborne surveillance. The idea that drives hyperspectral surveillance is that hyperspectral scanning draws information from such a large portion of the light spectrum that any given object should have a unique spectral signature in at least a few of the many bands that are scanned. The SEALs from NSWDG who killed Osama bin Laden in May 2011 used this technology while conducting the raid (Operation Neptune's Spear) on Osama bin Laden's compound in Abbottabad, Pakistan. Hyperspectral imaging has also shown potential to be used in facial recognition purposes. Facial recognition algorithms using hyperspectral imaging have been shown to perform better than algorithms using traditional imaging.

Traditionally, commercially available thermal infrared hyperspectral imaging systems have needed liquid nitrogen or helium cooling, which has made them impractical for most surveillance applications. In 2010, Specim introduced a thermal infrared hyperspectral camera that can be used for outdoor surveillance and UAV applications without an external light source such as the sun or the moon.

Astronomy

In astronomy, hyperspectral imaging is used to determine a spatially-resolved spectral image. Since a spectrum is an important diagnostic, having a spectrum for each pixel allows more science cases to be addressed. In astronomy, this technique is commonly referred to as integral field spectroscopy, and examples of this technique include FLAMES and SINFONI on the Very Large Telescope, but also the Advanced CCD Imaging Spectrometer on Chandra X-ray Observatory uses this technique.

Chemical Imaging

Remote chemical imaging of a simultaneous release of SF_6 and NH_3 at 1.5km using the Telops Hyper-Cam imaging spectrometer.

Soldiers can be exposed to a wide variety of chemical hazards. These threats are mostly invisible but detectable by hyperspectral imaging technology. The Telops Hyper-Cam, introduced in 2005, has demonstrated this at distances up to 5 km.

Environment

Top panel: Contour map of the time-averaged spectral radiance at 2078 cm^{-1} corresponding to a CO_2 emission line. Bottom panel: Contour map of the spectral radiance at 2580 cm^{-1} corresponding to continuum emission from particulates in the plume. The translucent gray rectangle indicates the position of the stack. The horizontal line at row 12 between columns 64^{-128} indicate the pixels used to estimate the background spectrum.

Most countries require continuous monitoring of emissions produced by coal and oil-fired power plants, municipal and hazardous waste incinerators, cement plants, as well as many other types of industrial sources. This monitoring is usually performed using extractive sampling systems coupled with infrared spectroscopy techniques. Some recent standoff measurements performed allowed the evaluation of the air quality but not many remote independent methods allow for low uncertainty measurements.

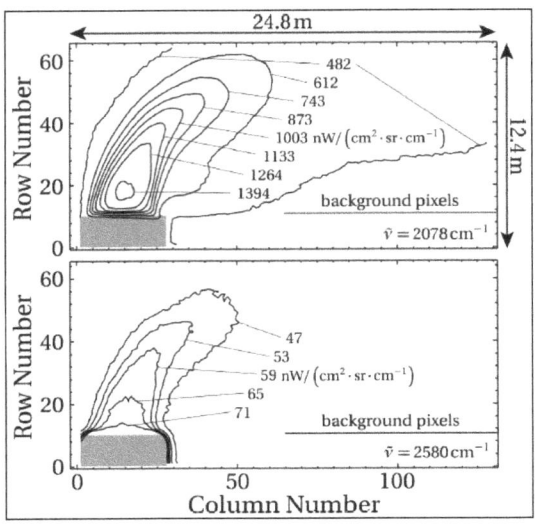

Advantages and Disadvantages

The primary advantage to hyperspectral imaging is that, because an entire spectrum is acquired at each point, the operator needs no prior knowledge of the sample, and post-processing allows all available information from the dataset to be mined. Hyperspectral imaging can also take advantage of the spatial relationships among the different spectra in a neighbourhood, allowing more elaborate spectral-spatial models for a more accurate segmentation and classification of the image.

The primary disadvantages are cost and complexity. Fast computers, sensitive detectors, and large data storage capacities are needed for analyzing hyperspectral data. Significant data storage capacity is necessary since hyperspectral cubes are large, multidimensional datasets, potentially exceeding hundreds of megabytes. All of these factors greatly increase the cost of acquiring and processing hyperspectral data. Also, one of the hurdles researchers have had to face is finding ways to program hyperspectral satellites to sort through data on their own and transmit only the most important images, as both transmission and storage of that much data could prove difficult and costly. As a relatively new analytical technique, the full potential of hyperspectral imaging has not yet been realized.

Remote Sensing in Geology

Remote sensing in geology is remote sensing used in the geological sciences as a data acquisition method complementary to field observation, because it allows mapping of geological characteristics of regions without physical contact with the areas being explored. About one-fourth of the Earth's total surface area is exposed land where information is ready to be extracted from detailed earth observation via remote sensing Remote sensing is conducted via detection of electromagnetic radiation by sensors. The radiation can be naturally sourced (passive remote sensing), or produced by machines (active remote sensing) and reflected off of the Earth surface. The electromagnetic radiation acts as an information carrier for two main variables. First, the intensities of reflectance at different wavelengths are detected, and plotted on a spectral reflectance curve. This spectral fingerprint is governed by the physio-chemical properties of the surface of the target object and therefore helps mineral identification and hence geological mapping, for example by hyperspectral imaging. Second, the two-way travel time of radiation from and back to the sensor can calculate the distance in active remote sensing systems, for example, Interferometric synthetic-aperture radar. This helps geomorphological studies of ground motion, and thus can illuminate deformations associated with landslides, earthquakes, etc.

Richat Structure by Shuttle Radar Topography Mission (SRTM). Instead of being a meteorite impact, the landform is more likely to be a collapsed dome fold structure.

Remote sensing data can help studies involving geological mapping, geological hazards and economic geology (i.e., exploration for minerals, petroleum, etc.). These geological studies commonly employ a multitude of tools classified according to short to long wavelengths of the electromagnetic radiation which various instruments are sensitive to. Shorter wavelengths are generally useful for site characterization up to mineralogical scale, while longer wavelengths reveal larger scale surface information, e.g. regional thermal anomalies, surface roughness, etc. Such techniques are particularly beneficial for exploration of inaccessible areas, and planets other than Earth. Remote sensing of proxies for geology, such as soils and vegetation that preferentially grows above different types of rocks, can also help infer the underlying geological patterns. Remote sensing data is often visualized using Geographical Information System (GIS) tools. Such tools permit a range of quantitative analyses, such as using different wavelengths of collected data sets in various Red-Green-Blue configurations to produce false color imagery to reveal key features. Thus, image processing is an important step to decipher parameters from the collected image and to extract information.

In remote sensing, the electromagnetic radiation acts as the information carrier, with a distance of tens to thousands of kilometers distance between the sensor and the target. Proximal Sensing is a similar idea but often refer to laboratory and field measurements, instead of images showing a large spatial extent Geophysical methods, for instance Sonar and acoustic methods, shares similar properties with remote sensing but electromagnetic wave is not the sole medium. Geotechnical instrumentations, for example piezometer, tiltmeter and Global Positioning System (GPS), on the other hand, often refer to instruments installed to measure discrete point data, compared to imagery in remote sensing. A suitable sensor sensitive to the particular wavelength region, according to the designated use, is selected and employed to collect the electromagnetic wave reflected or emitted from the target object.

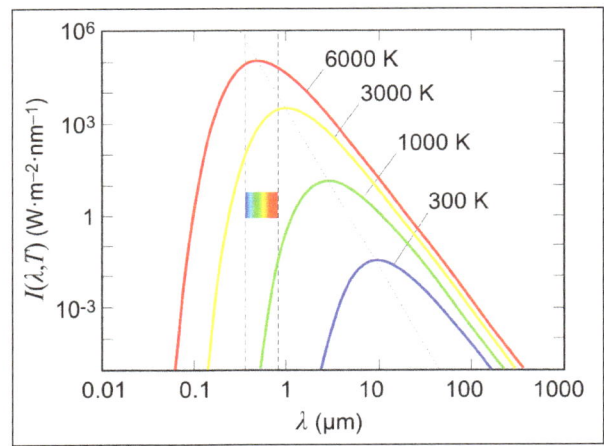

Thermal emission according to Planck's Law. The Sun is approximately 6000K
in surface temperature and the emission peaks at visible light. The Earth, approximated
to 300K also emit non-visible radiation.

Working Principles

In remote sensing, two main variables are measured in a typical remote sensing system: the radiance (or intensity) and time of arrival for active systems. The radiance (i.e. returning signal intensity) versus wavelength is plotted to a spectral reflectance curve. As a point to note, the data collected is a blend of both reflection of solar radiation and emission (according to Planck's law) from the object for visible and near infrared (VNIR) region. The thermal infrared (TIR) region measures mainly emission while microwave region record backscattering portion of reflection. The radiance is determined by radiation-matter interactions, which is governed by the physio-chemical properties of the target object. Prominent absorptions in specific wavelength shown on the spectral reflectance curve are the fingerprints for identification in spectroscopy. The two-way travel time of the radiation could infer the distance as the speed is approximately equal to the speed of light, roughly 3×10^8 m/s. This allows application in ranging in light detection and ranging (LiDAR) and Radio detection and ranging (Radar) etc.

Since the sensors are looking through the atmosphere to reach the target, there are atmospheric absorption. Three main atmospheric windows, which allow penetration of radiation, can be identified. They are 0.4–3 micro-meters (Visible and Near-Infrared (VNIR)), 3–14 micro-meters (Thermal Infrared TIR) and few millimeters to meters (microwave). Camera in everyday life is a passive imaging system in VNIR wavelength region. A simple classification of prevailing remote sensing instruments in geology, modified from Rees in accordance with context of this page.

Wavelength Range	Wavelength	Sensitive to	Passive	Active Systems (Ranging or Imaging)
VNIR	0.4-3 micro-meters	Intra-atomic electronic transitions	Spectroscopy [spectrometer]; Aerial photography/ Photogrammetry [camera]	[LiDAR]
TIR	3–14 micro-meters	Inter-atomic bond strength in molecules	[TIR Imager]	/
Microwave	few millimeters to meters (microwave)	temperature, terrain roughness, particle size	/	Synthetic Aperture Radar/ InSAR [Radar]

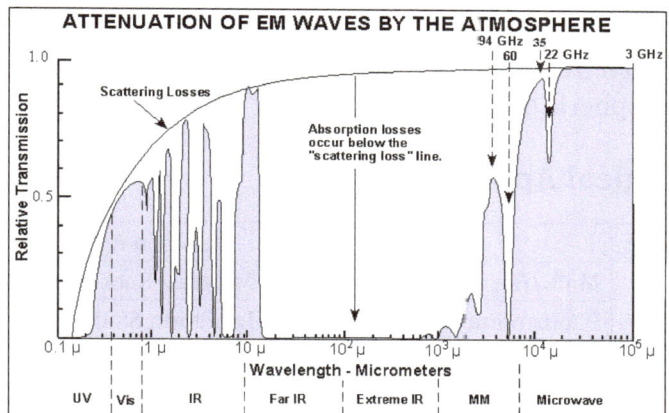

Relative transmission of radiation with respect to wavelength. There are 3 atmospheric windows (VNIR, TIR and Microwave) allowing radiation to penetrate through the atmosphere without prominent absorption. Some corrections are still needed to remove the atmospheric attenuation.

Carrying Platform

The sensor could be spaceborne (carried by satellite), airborne (carried by aircraft, or most recently Unmanned Aerial Vehicle (UAV)) or ground-based (sometimes called proximal sensing). Data acquired from higher elevation captures a larger field of view/ spatial coverage, but the resolutions are often lower. Prior mission planning regarding flight path, weight load, carrying sensor etc. have to be done before deployment. The resolution requirement is often high in geological studies, hence airborne and ground-based systems prevail in surveying.

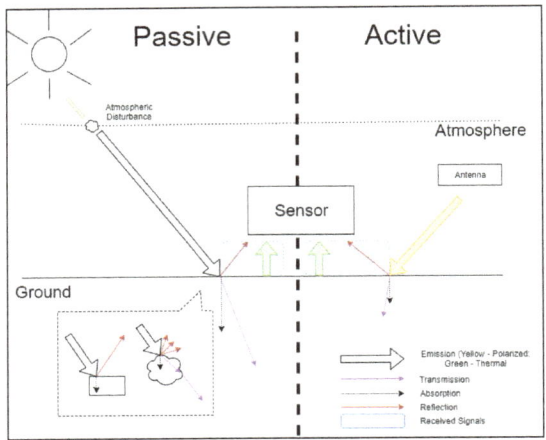

Schematic drawing of Passive (left) and Active (right) remote sensing. The radiation-matter interaction in microscopic scale (absorption, transmission and reflection) is depicted in the left bottom corner speech box. The relative proportion is governed by physio-chemical properties of the material. Planar surface promotes specular reflection while rough surface gives a diffused reflection.The sensor detects (blue box) reflection of solar radiation from the target in passive remote sensing, while active remote sensing systems illuminate the target and detect the reflection. Both passive and active receive naturally emitted thermal radiation emitted according to Planck's Law. They are also subject to atmospheric disturbance.

Common Geological Application

Wavelength Range	Tools	Common Applications in Geology
VNIR	Multi/Hyper-spectral Imaging	Mineral/ Rock Identification
	Photogrammetry	Landform Studies
	LiDAR	Geodetic Survey
TIR	Thermal infrared Imaging	Thermal Anomalies
Microwave	Synthetic Aperture Radar	Displacement Time-series

Advantages and Limitations

The main advantage of employing remote sensing to deal with geological problem is that it provides direct information on the surface cover using a synoptic coverage or sometimes stereoscopic view. Hence the big picture of kinematics could be better appreciated. It also reduces the burden of field work required for the area through synoptic studies of the area of interest. The spectral vision allows identification of rock attributes for surficial mapping. The resolution however controls the accuracy.

There is a trade-off between spatial resolution and spectral resolution. Since the intensity of incident ray is fixed, for a higher spectral resolution, it is expected to have a lower spatial resolution (one pixel represent larger area) to maintain an up-to-standard signal-to-noise ratio for analysis. Also, the data volume for transmission is limited due to signal engineering problem. One can never obtain data with maximum resolutions in all spatial, spectral and radiometric resolutions due to these constraints. The temporal resolution could be understood as both the revisiting frequency and the deformation duration. For instance an instantaneous landslide or sinkhole collapse could hardly be recorded without high speed camera, while relics could be imaged into time-series where the temporal change, for instance ice calving could be revealed.

Another deficiency is the inconsistent data acquisition method, plus their interpretation schemes. As a result, an ideal database is hardly feasible because of dynamic environmental conditions in different locality. Instead, repeated reconnaissance is suggested for studying a specific area.

Field observation and reconnaissance remains irreplaceable and shall never be taken over completely by remote sensing because field data greatly support remote sensing data interpretation. Remote sensing should better be considered to be complementary which aims to provide instantaneous views of different scale, perspective or spectral vision. Subsurface mapping by Geophysical survey and simulation models still play an important role in the 3 dimensional characterization of soil underground. A word of caution is that there is no such an 'ideal' sensor capable or optimized to study everything. It is often up to scientist's preference and experience to pick which dataset and extract information. For instance aerial photographs could be more sensible in cloud-free area, but otherwise radar may perform better for overcast weather.

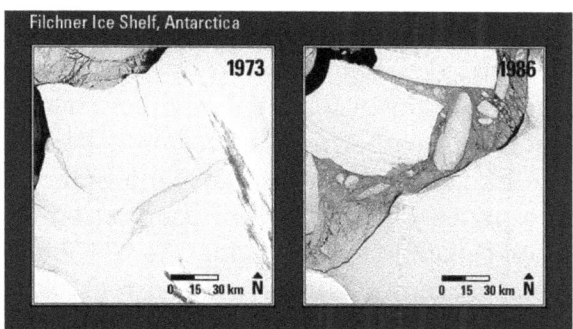

The break off of Filchner Ice Shelf, Antarctica. The near-infrared reflectance image distinguishes water from ice – Landsat.

Geological Mapping

Spectral Features

Remote sensing can aid surficial geological mapping and landform characterization.

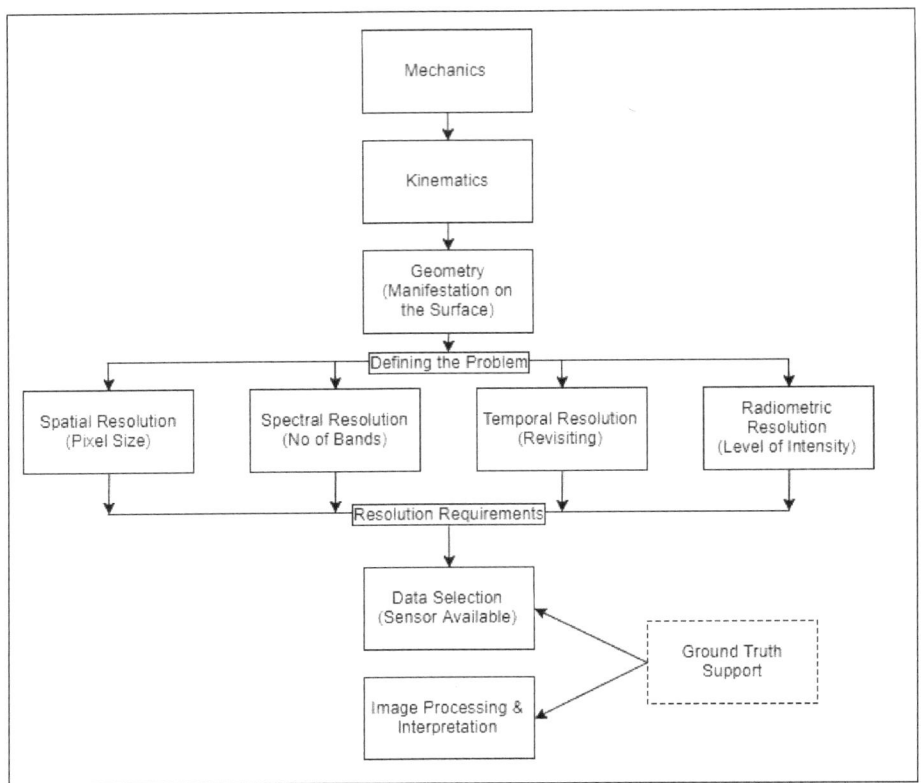

A typical workflow for tackling geological problem, starting from defining the problem down to data selection and interpretation.

The visible and near infrared (VNIR) and thermal infrared (TIR) are sensitive to intra-atomic electronic transitions and inter-atomic bond strength respectively can help mineral and rock identifications.The instrument in use is called spectroradiometer in lab and imaging spectrometer or multi-/ hyper-spectral scanner as imaging remote sensors. Provided that the land is not obscured by dense vegetation, some characteristics of superficial soil (the unconsolidated sedimentary materials covering the land as surficial deposits from weathering and erosion of bedrock) may be measured with a penetration depth into air-soil interface of about half of wavelength used (e.g. green light (~0.55 micro-meters) gives depth of penetration into ~0.275 micro-meters). Hence most remote sensing systems using the VNIR wavelength region give characteristics of surface soil, or sometimes exposed rock. Another parameter controlling the overall reflectance is surface roughness. The same surface can appear rough in VNIR may appear smooth in microwave, similar to what we perceive when we use a meter rule to measure roughness where surface fluctuation are in cm-scale. As grain size decreases, surface roughness increases and hence overall reflectance increases as diffuse reflection, instead of specular reflection, dominates. Specular reflection by smooth surface, for example calm water, gives little backscattering and hence appear dark. As an example, ice is mostly transparent in a large piece but becomes highly reflective when smashed into small grains.

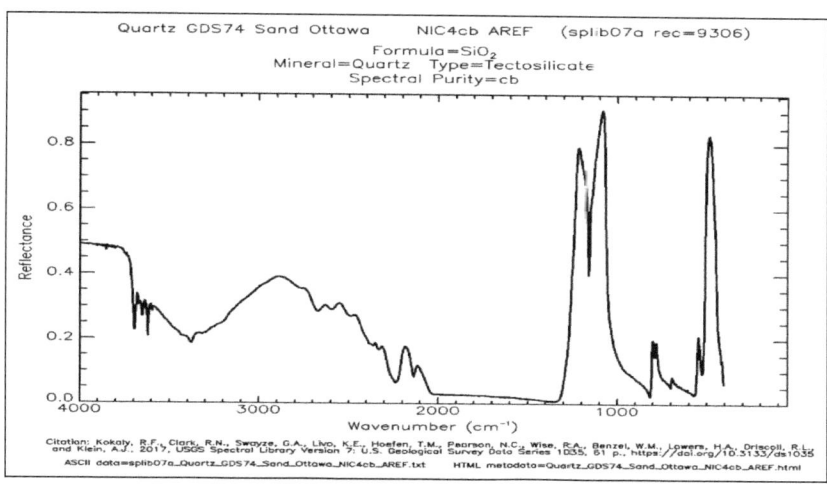

An example of a mineral quartz spectral reflectance curve.

Mineral and Rock

In lithological composition studies, lab and field spectroscopy of proximal and re-mote sensing can help. The spectral reflectance data from imaging spectrometry employing short wavelength, for example form Airborne visible/infrared imaging spectrometer (AVIRIS), provide chemical properties of the target object. For in-stance, the iron content, which is indicative of the soil fertility and the age of sedi-ment, could be approximated. For soil with high iron oxide, which is red in colour, should give higher reflectance in red wavelength portion and diminishes in blue and green. There may also be absorption at 850-900 nm. The Redness Index and absorption area in 550 nm in spectral reflectance curve are examples to quantify iron content in soil.

To identify mineral, available spectral reflectance libraries, for instance the USGS Spectral Library, summarize diagnostic absorption bands for many materials not lim-ited to rocks and minerals. This helps create a mineral map to identify the type of min-eral sharing similar spectra, with minimal in situ field work. The mineralogy is identi-fied by matching collected sample with spectral libraries by statistical method, such as partial least squares regression. In addition to high signal-to-noise ratio (>40:1), a fine spatial resolution, which limits the number of elements inside one single pixel, also promotes decision accuracy. There are also digital subpixel spectral unmixing tools available. The USGS Tetracorder which applies multiple algorithms to one spectral data with respect to the spectral library is sensitive and give promising results. The different approaches are summarized and classified in literature but unfortunately there is no universal recipe for mineral identification.

For rocks, be they igneous, sedimentary or metamorphic, most of their diagnostic spec-tral characteristics of mineralogy are present in longer wavelength (SWIR and TIR), which is for example present in the ASTER mission. This is due to the sensitivity of vibrational bands of longer wavelength. As opposed to the automatic statistical inter-pretation mentioned above for minerals, it is more advisable to adopt visual interpre-tation for identification of rock because the surficial alteration of rock may exhibit very different spectral responses.

Several indices are proposed for rock type identification, such as Quartz Index, Carbon-ate Index and Mafic Index, where Di is the data of the i-th band in ASTER.

- Carbonate Index (CI): D_{13}/D_{14}.

- Quartz Index (QI): $D_{11}*D_{11} / D_{10}*D_{12}$.

- Mafic Index (MI): D_{12}/D_{13}.

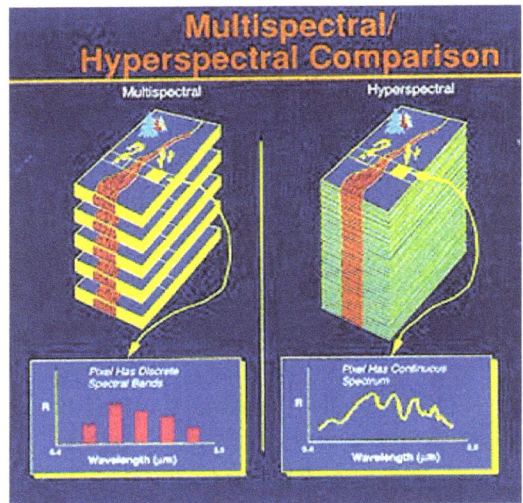

Hyperspectral imaging gives high spectral resolution, but as a trade-off the spatial and radiometric resolutions are lower.

Soil

Surficial soil is a good proxy to the geology underneath. Some of the properties of soil, alongside lithology mentioned above, are retrievable in remote sensing data, for in-stance Landsat ETM+, to develop the soil horizon and therefore aid its classification.

Soil Texture and Moisture Content

The amount of moisture within soil particles is governed by the particle size and soil texture as the interstitial space may be filled with air for dry soil and water for saturated soil. Essentially, the finer the grain size, the higher capability to hold moisture. As men-tioned above, wetter soil is brighter than dry soil in radar image. For short wavelength VNIR region, in the same theory, clayey surface with smaller grain size promoting more backscattering should give higher spectral response. However, the higher soil moisture and organic contents makes clay darker in images, compared to silty and sandy soil cover after precipitation. With regard to VNIR region, as the moisture content increases, more prominent absorption (at 1.4, 1.9, 2.7 micrometers, and sometimes at 1.7 for hydroxyl absorption band) take place. On the other hand, radar is sensitive to one more factor: dielectric constant. Since water has a high dielectric constant, it has high reflectivity and hence more backscattering takes place, i.e. appears brighter in radar images. Therefore, soil appears brighter with higher soil moisture content (with the presence of capillary water) but appears dark for flooded soil (specular reflection). Quantitatively, while soil texture is determined by statistical means of regression with calibration, scientists also developed a Soil Water Index (SWI) for long-term change detection. Another approach is surface energy balance model, which makes prediction on the actual evapotranspiration.

In short, the soil moisture overall reflectance could be tabulated.

Tools	Dry Soil	Wet Soil	Flooded Soil
Radar	Darker (energy penetrate into soil of low dielectric constant)	Brighter (water has high dielectric constant)	Very dark (specular reflection)
VNIR	Brighter (less absorption)	Darker (prominent water absorption)	Same as water (low penetration depth)

Soil Organic Carbon

Soil organic carbon is mainly derived from proximal sensing using mid infrared reflectance. A darker soil colour results from the saturated organic matter plus variable amount of black humic acid and soil moisture. The higher the amount of organic content in the soil, incident energy would be greatly absorbed and as a result lower reflectance is expected in general. The contrast in soil colour allows quantitative analysis of band depth analysis, principal component analysis and modeling.

Soil Salinity

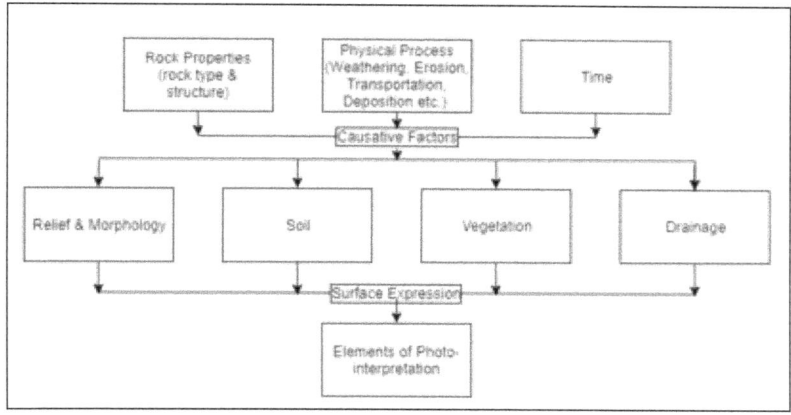

The surface manifestations of the geological kinematics
provide clues for photo interpretation.

Soil salinity is the result of insufficient rainwater precipitation which leads to accumulation of soluble salt in soil horizon. The spectral proximal sensing in VNIR is sensitive to absorption of water in hydrated evaporate minerals, most prominently at 505 nm, 920 nm, 1415 nm and 2205 nm. For even more saline soil, 680, 1180 nm and 1780 nm would also give lower reflectance (higher absorption) and higher reflectance at 2200 nm possibly due to the loss of crystallinity in clay minerals. The spectral curve also shows a decreasing overall slope from 800 nm to 1300 nm for higher salinity. The overall reflectance curve in all wavelength increases with higher salt concentration, while the compositions of salt would show some variations in absorption bands.

Geomorphology

3-dimensional geomorphological features arising from regional tectonics and formation mechanisms could also be understood from a perspective of small scale images showing a large area acquired in elevation. The topography of an area is often characterized by volcanic activity or orogenesis. These mountain building processes are determined by the stress-strain relation in accordance with rock types. They behave as elastic/ plastic/ fracturing deformations, in response to different kinetics. Remote sensing techniques provide evidence such as observed lineament, global scale mountain distribution, seismicity and volcanic activities to support crustal scale tectonics and geodynamics studies. Additional spectral information also helps. For example, the grain size differentiates snow and ice. Aside from a planar geological map with cross-sections, sometimes 3-dimensional view from stereo-photos or representation in Digital Elevation Model (DEM) could aid the visualization. In theory, LiDAR gives the best resolution up to cm grade while radar gives 10m grade because of its high sensitivity to small scale roughness. Oblique images could greatly enhance the third-dimension, but users should bear in mind the shadowing and distortions.

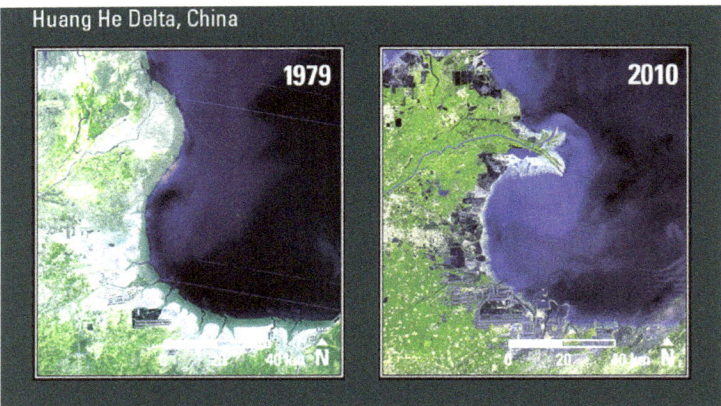

The deltaic landform at the mouth of HuangHe, China – Landsat.

Inaccessible Areas

Although field mapping is the most primary and preferable way to acquire ground truth, the method does not work when areas become inaccessible, for example the conditions are too dangerous or extreme. Sometimes political concerns bar scientists' entering. Remote sensing, on the other hand, provides information of the area of interest sending neither a man nor a sensor to the site.

Desert

Desert area is shaped by eolian processes and its landforms by erosion and deposition. The stereopairs of aerial photos provide three-dimensional visualization for the land feature while hyperspectral image provide grain scale information for grain size, sand composition etc. The images are often of high phototones in short wavelengths in VNIR corresponding to the scanty vegetation and little moisture. Another tool is radar, which has the capability to penetrate surficial sand, a material of low dielectric constant. This see-through characteristic, notably of the L-band (1.25 GHz) microwave of 1–2 m penetration, allows subsurface mapping and possibly identification of past aquifer. The paleohydrography in Sahara Desert and Gobi Desert is revealed and further studies using airborne P-band (435 MHz) for penetration of 5 m is proposed in future research.

Political Sensitive Area

Politics poses challenge to scientific research. One example is the Tibesti Mountains, which is remote and politically unstable in recent decades due to conflicts over uranium deposit underneath. The area, however, could possibly serve as one spectacular example of intracontinental volcanism or hotspot. Detailed studies of the area divided into Western, Central and Eastern Tibesti Volcanic Province shows no significant sign of spatially progressive volcanism, and hence it is unlikely to be a hotspot as the manifestation of Hawaiian or Galapagos Islands. More data and systematic field survey and studies, for instance geochemical and radiometric dating, are great topics to focus on in the near future. The Tibesti board swell dome is also studied as a regional river system to estimate the uplift rate.

Water Bodies

Water bodies, for instance ocean and glaciers, could be studied using remote sensing. Here are two examples for plankton and glacier mapping.

The bloom of photosynthesizing phytoplankton is an ecological proxy to favorable environmental conditions. Satellite remote sensing in VNIR wavelength region help locate sporadic event of change in ocean colour due to relative increase in related absorption in spectral curve. Different band math (e.g. band ratio algorithms and spectral band difference) are developed to cater coastal and open water, and even some specific types of bloom (e.g. Coccolithophore blooms and Trichodesmium blooms). The capability of real-time monitoring by combined use of long term satellite data allows better insight into the ocean dynamics.

The mapping of glaciers is facilitated by remote sensing in contrast with carrying heavy equipment to dangerous icy land. Some notable applications include mapping of clean-ice and debris-covered glaciers, glacier fluctuation records, mass balance and volume change studies to aid generating topographic map and quantitative analysis. Similarly, automated approach using band math and DEM calculations using high resolution data is requisite to look into the glacial variations due to dynamic environmental conditions.

Geologic Hazards

Geological hazards cause casualties and serious damage to properties. While it is almost impossible to prevent naturally occurring disasters, their impact could be reduced and minimized with proper prior risk assessment and planning.

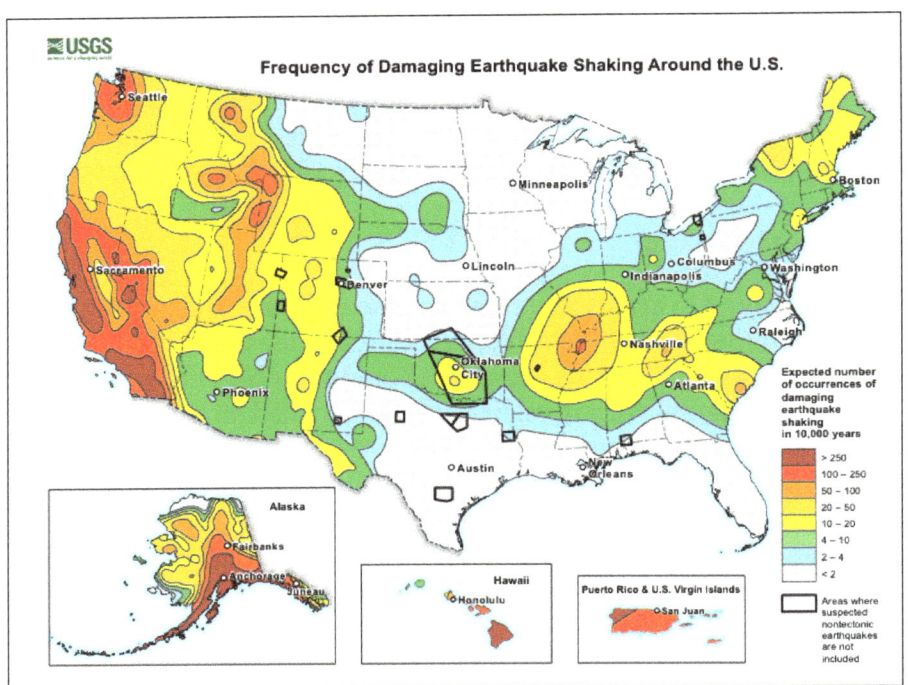

U.S. Seismic Hazard Maps 2014.

Earthquakes

Earthquakes manifest itself in movement of earth surface. Remote sensing can also help earthquake studies by two aspects. One is to better understand the local ground condition. For instance some soil type, which is prone to liquefaction (e.g. saturated loose alluvial material), do more damage under vibration and therefore earthquake hazard zoning may help in reducing property loss. Another one is to locate historical earthquakes in neotectonism (past 11000 years) and analysis its spatial distribution, and hence fault zones with structural ruptures are mapped for further investigations. From a geodetic perspective, the radar technique (SAR Interferometry, also called In-SAR) provides land displacement measurement up to cm scale. SAR interferometry is a technology utilizing multiple SAR images recording the backscattering of microwave signal. The returning signal can be used to estimate the distance between ground and satellite. When two images are obtained at the same point but at different time, some pixels showing delayed returns reveal displacement, assuming no ground change. A displacement map (interferogram) is generated to visualize the changes with a precision up to a half of the wavelength i.e. cm grade. Another similar technique is Global Positioning System (GPS), which records the displacement with time of discrete points through trilateration of microwave GPS satellite signals. The same idea and principle to measure ground displacement could also be extended to other natural hazards monitoring, such as volcanisms, landslides and avalanches. The mid-IR thermal (11–12 micrometer) satellite images have shown some thermal fields in geological active areas, such as lineation and fault systems. Aside from these long-lived thermal fields, there are some positive thermal anomalies of 3–4 °C on land surface or around −5 °C for sea water in earthquake epicenter areas. The contrast appears 7–14 days prior to the earth movement. Though the observation is supported by laboratory experiments, the possible causes of these differences are still debatable.

Tsunami

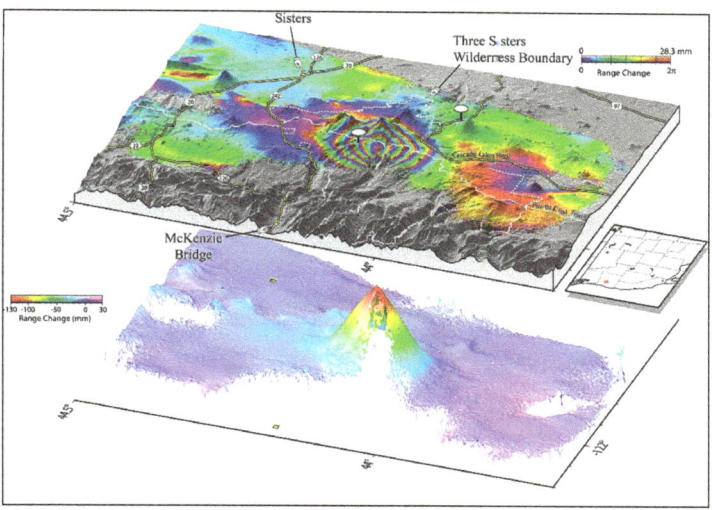

Mangrove offers protection against tsunami and storms from sweeping away inland areas because the mangrove fringes were damaged and took all the energy. Remote sensing of mangrove and vegetation as a natural barrier to manage risks therefore becomes a hot topic. The recent advancement and development is highly anticipated in the near future, especially as hyperspectral imaging system and very high resolution (up to sub meter grade) satellite images prevails. New classification schemes distinguishing species from composition could be developed for environmental studies. Estimation leaf area, canopy height, biomass and productivity could also be improved with the emergence of existing and forthcoming sensors and algorithms. Tsunami-induced inundation leads to change in coastal area, which could be quantified by remote sensing. Split-based approach to divide large images into subimages for further analysis by redefining change detection threshold have reduced computation time and have shown to be consistent with manual mapping of affected areas.

An interferogram showing ground movement in the Three Sisters Wilderness, where eruption occurred 1500 years ago. Each colour contour represents an equal amount of uplift, which is possibly caused by magma accumulation at about 7 km depth. The uplift is about 130mm with lateral extent of 20 km. The white thumbtacks are GPS stations.

References

- "Studies on Hyperspectral Face Recognition in Visible Spectrum With Feature Band Selection - IEEE Journals & Magazine". Ieeexplore.ieee.org. Retrieved 2019-01-06

- Remote-sensing-for-natural-or-man-made-disasters-and-environmental-changes, environmental-applications-of-remote-sensing, books: intechopen.com, Retrieved 6 April, 2020

- Lu, G; Fei, B (January 2014). "Medical Hyperspectral Imaging: a review". Journal of Biomedical Optics. 19 (1): 10901. Bibcode:2014JBO....19a0901l. Doi:10.1117/1.JBO.19.1.010901. PMC 3895860. PMID 24441941

- Remote-sensing-applications-in-agriculture, remote-sensing: grindgis.com, Retrieved 7 May, 2020

- Hans Grahn; Paul Geladi (27 September 2007). Techniques and Applications of Hyperspectral Image Analysis. John Wiley & Sons. ISBN 978-0-470-01087-7

- Remote-sensing-applications, remote-sensing: grindgis.com, Retrieved 8 June, 2020

- Fernández Pierna, J.A., et al., 'Combination of Support Vector Machines (SVM) and Near Infrared (NIR) imaging spectroscopy for the detection of meat and bone meat (MBM) in compound feeds' Journal of Chemometrics 18 (2004) 341-349

Permissions

We would like to thank the editorial team for lending their expertise to make the book truly unique. They have played a crucial role in the development of this book. Without their invaluable contributions this book wouldn't have been possible. They have made vital efforts to compile up to date information on the varied aspects of this subject to make this book a valuable addition to the collection of many professionals and students.

This book was conceptualized with the vision of imparting up-to-date and integrated information in this field. To ensure the same, a matchless editorial board was set up. Every individual on the board went through rigorous rounds of assessment to prove their worth. After which they invested a large part of their time researching and compiling the most relevant data for our readers.

The editorial board has been involved in producing this book since its inception. They have spent rigorous hours researching and exploring the diverse topics which have resulted in the successful publishing of this book. They have passed on their knowledge of decades through this book. To expedite this challenging task, the publisher supported the team at every step. A small team of assistant editors was also appointed to further simplify the editing procedure and attain best results for the readers.

Apart from the editorial board, the designing team has also invested a significant amount of their time in understanding the subject and creating the most relevant covers. They scrutinized every image to scout for the most suitable representation of the subject and create an appropriate cover for the book.

The publishing team has been an ardent support to the editorial, designing and production team. Their endless efforts to recruit the best for this project, has resulted in the accomplishment of this book. They are a veteran in the field of academics and their pool of knowledge is as vast as their experience in printing. Their expertise and guidance has proved useful at every step. Their uncompromising quality standards have made this book an exceptional effort. Their encouragement from time to time has been an inspiration for everyone.

The publisher and the editorial board hope that this book will prove to be a valuable piece of knowledge for students, practitioners and scholars across the globe.

Index

Lightning Source UK Ltd.
Milton Keynes UK
UKHW050329310822
408102UK00002B/31